# Simulation Foundations, Methods and Applications

**Series Editor**

Andreas Tolk, The MITRE Corporation, McLean, VA, USA

**Advisory Editors**

Roy E. Crosbie, California State University, Chico, CA, USA

Tony Jakeman, Australian National University, Canberra, ACT, Australia

Axel Lehmann, Universität der Bundeswehr München, Neubiberg, Germany

Stewart Robinson, Business School, Newcastle University, Newcastle upon Tyne, UK

Claudia Szabo, School of Computer Science, University of Adelaide, Adelaide, Australia

Mamadou Kaba Traoré⬤, IMS UMR 5218, University of Bordeaux, Talence, France

Bernard P. Zeigler, University of Arizona, Tucson, AZ, USA

Lin Zhang, Beihang Unviversity, Beijing, China

Andreas Tolk, Old Dominion University, Norfolk, VA, USA

Rodrigo Castro, Facultad de Ciencias Exactas y Naturales, Departamento de Computación, Universidad de Buenos Aires, Ciudad Autónoma de Buenos Aires, Argentina

Sanja Lazarova-Molnar, Institute of Applied Informatics and Formal Description Methods, Karlsruhe Institute of Technology, Karlsruhe, Germany

The modelling and simulation community extends over a range of diverse disciplines and this landscape continues to expand at an impressive rate. Modelling and simulation is fundamentally a computational tool which has an established record of significantly enhancing the understanding of dynamic system behaviour on one hand, and the system design process on the other. Its relevance is unconstrained by discipline boundaries. Furthermore, the ever-increasing availability of computational power makes feasible applications that were previously beyond consideration.

*Simulation Foundations, Methods and Applications* hosts high-quality contributions that address the various facets of the modelling and simulation enterprise. These range from fundamental concepts that are strengthening the foundation of the discipline to the exploration of advances and emerging developments in the expanding landscape of application areas. The underlying intent is to facilitate and promote the sharing of creative ideas across discipline boundaries. The readership will include senior undergraduate and graduate students, modelling and simulation professionals and research workers.

Inasmuch as a model development phase is a prerequisite for any simulation study, there is an expectation that modelling issues will be appropriately addressed in each presentation. Incorporation of case studies and simulation results will be strongly encouraged.

Titles can span a variety of product types, including but not exclusively, textbooks, expository monographs, contributed volumes, research monographs, professional texts, guidebooks and other references.

These books will appeal, varyingly, to senior undergraduate and graduate students, and researchers in any of a host of disciplines where modelling and simulation has become (or is becoming) a basic problem-solving tool. Some titles will also directly appeal to modelling and simulation professionals and practitioners.

Stanislaw Raczynski

# Models for Research and Understanding

## Exploring Dynamic Systems, Unconventional Approaches, and Applications

 Springer

Stanislaw Raczynski [ID]
Facultad de Ingenieria
Universidad Panamericana
Mexico City, Mexico

ISSN 2195-2817 ISSN 2195-2825 (electronic)
Simulation Foundations, Methods and Applications
ISBN 978-3-031-11928-6 ISBN 978-3-031-11926-2 (eBook)
https://doi.org/10.1007/978-3-031-11926-2

This Springer imprint is published by the registered company Springer Nature Switzerland AG
The registered company address is: Gewerbestrasse 11, 6330 Cham, Switzerland

# Preface

In this book, you will find topics in modeling and simulation, both conventional, as well as suggestions about new tools and somewhat wider insight on modeling. Most of the chapters of the book may be used in didactic work, as a textbook or supplementary material. In the text, the student will find explanations of diverse kind of modeling tasks, and examples of models of various types. Ten out of 12 chapters contain also sections of questions and answers that focus on the most essential aspects of the presented modeling methodology.

Models are used by thinking beings, consciously or sub-consciously, to manage their behavior and respond to the events in the environment. The most perfect and huge collection of models is contained in our brains. During years of learning, the humans, as well as most of the animals, created models of their own bodies and objects from the environment, to be able to take decisions and control their behavior. The human brain is increasingly complicated and capable of creating and storing models not only inside it, but also in other media. We create models in the form of written descriptions, drawings, mathematical expressions, or logical structures, and store them for further use.

The appearance of computers provided a new, powerful tool for model storage and processing. In this book, we discuss models that can be used in computer simulation. This kind of modeling is closely connected with simulation on analog and digital machines. The results of creating and using models help us to take decisions, predict, and/or understand the world where we live. Most of the things we observe are still too complex to be understood. The knowledge accumulated in models helps us in taking correct actions, grow, understand, and survive.

In general, models help us to understand how complex systems work. Models of complex systems need not be complex or huge. I remember that in the early 1960s, when we worked on 100kHz CPU, 32kB memory machines, a friend of mine asked me to solve an ordinary differential equation on a computer. The equation was written in one line, with rather simple, but non-trivial formula. The solution was not easy, but finally the program worked. I asked him: *This is the model of what?* The answer was simple: *This is the model of the Universe.*

On the other hand, there are huge models that provide little help in complex system understanding. In Chap. 1, there is a remark on the model of the economy of the whole Soviet imperium of 1970s, developed by the Central Economic Mathematical Institute of the URSS. As the authors agreed, the model resulted to be too large and complicated to be simulated and interpret.

Here, you also will find here several topics closely related to model building, like stability of linear and non-linear models, model validity, credibility, new approach to sensitivity, and some more generalized insight, based on the theory of categories. New tools for model creation are proposed, namely, the differential inclusions and reachable sets calculation. For the discrete event models, some problems of simultaneous events are discussed. A new methodology of the *semi-discrete events* with finite-time duration is suggested to eliminate the ambiguity that arises in strictly discrete models with simultaneous events.

This book does not pretend to describe and discuss all this huge research and experience gathered during many decades. We try to give an overview of the most important modeling issues. The tenet of the book is to show also some non-typical and unconventional kind of models and their applications, like models based on differential inclusions and semi-discrete event models.

Looking at the annals of modeling and simulation activity, it can be seen that most of the simulationists and model-makers follow the same modeling paradigms for more than six decades. There exists a *strange conviction that everything that is continuous can be modeled by differential equations*, and the only tool to simulate events is the discrete event simulation. Here, we show some alternative ways to construct and treat models, like differential inclusions, reachable set calculation, and semi-discrete event simulation. Note that the concepts as *functional sensitivity* (Chap. 4) is new and original. Note also that the images of reachable sets of dynamic systems shown here (Chaps. 3–5) have never been provided before in other sources. We also discuss a somewhat non-conventional way of model description that uses the language of the theory of categories.

It should be emphasized that **this book treats about dynamic models**. We do not discuss static models that can be treated by algebraic methods or static modeling and optimization. This is because **the real world is dynamic**. Everything is changing, fluctuating, or oscillating. The *time* variable appears in all models we present.

We do not discuss here models being developed in some specific fields of research like models of neural networks, weather forecasting, human brain, missile dynamics, human immune system, games, electromagnetic wave propagation, human DNA, urban development, and many others. If we do this, the book would have several thousands of pages. We also omit the recall of probability distributions used in computer simulation because such overview comes in any of hundreds of the books on computer simulation. When talking about modeling and simulation, the question arises: Does the model creator or the simulationist must be a mathematician? My point is that he/she doesn't. However, some basic knowledge of mathematics is required (at least for the readers of this book). This may prevent the researcher from making serious errors like using the Poisson distribution for between-arrival time intervals of the Poisson arrival process, treating the affine function as lineal, or calculating statistics over a data set of five points.

We also omit issues and examples of models already published by the author in Springer Nature series. These are: interactions between hierarchical structures, dynamics of terror and anti-terror organizations, organization growth and decay models, human migration, extended prey–predator model, and others, discussed in S. Raczynski, "Interacting complexities of herds and social organizations." The models of catastrophes, unexpected behavior patterns in artificial societies, catastrophes in stock market, epidemics, growth of organisms, work, salary, and motivation are described in "Catastrophes and unexpected behavior patterns in complex artificial societies," by the same author.

In Chap. 1, there is an overview about the general concepts of modeling. The continuous and discrete event models are shortly discussed. The main concepts of model validity and mathematical and graphical representation of models are considered. Chapter 2 is dedicated to continuous models. The conventional approach to continuous modeling and simulation is described.

In Chap. 3, the reader can find the discussion about the concepts and applications of models based on the *differential inclusions*. This tool is useful in the analysis of model uncertainty and sensitivity. The new method named *functional sensitivity* is proposed. More detailed examples of application of the concept of functional sensitivity can be found in Chap. 4.

Chapter 5 describes a model of airplane flight control and reachable sets.

An application to Pontryagin's Maximum Principle to market optimization can be found in Chap. 6. This chapter shows the connection between modeling and optimization. An example of optimization technique based on the theory of optimal control is given. In this example, the reader can see the optimization of the dynamic marketing strategies.

An overview of discrete event models and simulation is provided in Chap. 7, including the Discrete Event Specification Formalism and distributed simulation.

In Chapter 8 contains an example of a model of self-organization in a population of living beings, and interactions between organizations and the environment.

Chapter 9 contains certain non-conventional approach to the general theory of models. A *metrics in the space of models* is defined. In this chapter, there is a proposal of the new concept of *semi-discrete event* simulation.

In Chap. 10, some examples of application of the language of the theory of categories in model specifications are discussed.

The last two chapters do not include question and answer sections. These chapters present examples of somewhat unconventional approach, and more abstract modeling tasks. The main purpose of these chapters is to show models that perhaps have little practical applications, but offer a new and cognitive topics.

Chapter 11 deals with the concept of time variable and the time instant. The proposed model of the "time arrow" includes the concept of a *fuzzy time instant*. Instead of considering the events that occur in given, sharp moments of time we treat the time instant as an interval with non-zero length, where events "occur" gradually. Some issues related to the causality principle are discussed.

In Chap. 12, the reader can find a new approach to models related to the real time, the model reversibility and uncertainty of the future. An abstract, but perhaps interesting model of the universe encapsulated in a ball of finite size is given. The idea of the universe encapsulated in a multi-dimensional ball may appear to be very abstract. However, this idea may provoke the reader to ask "And what is outside the ball?" and "Is it possible to out of it?" In the last question, we supposes the possibility to travel "behind the infinity." Moreover, we can add the fourth dimension to the ball (the *time*), and then add the fifth and more dimensions. This is a non-science-fiction remark on possible five- and multi-dimensional universe.

Mexico City, Mexico                                                                 Stanislaw Raczynski

# Acknowledgements

I would like to express my gratitude to the Editors of the journals listed below for the permission to use the updated versions of my articles, as follows:

Semi-discrete events and models in categorical language. International Journal of Simulation Modelling, 2012; 11, (2), 77–88;

Uncertainty, dualism and inverse reachable sets. International Journal of Simulation Modelling, 2011; 10, (1), 38–45, ISBN/ISSN: 1726–4529;

and

Simulating self-organization and interference between certain hierarchical structures. Nonlinear Dynamics Psychology and Life Sciences, Human Sciences Press, 2014; 18, (4), 419–434, ISBN/ISSN: 1090-0578.

Stanislaw Raczynski

# Contents

**1    Concept of Model** ............................................... 1
   1.1    Introduction: General Remarks ........................... 1
   1.2    The System ............................................... 3
   1.3    Mathematical Models ...................................... 3
           1.3.1    Kinds of Mathematical Models ................... 5
           1.3.2    Models of Economic Growth ..................... 6
           1.3.3    Models in Public Health and Epidemics ......... 7
           1.3.4    Graphical Representations of Continuous Models .... 9
           1.3.5    Computational Tractability ..................... 10
   1.4    Discrete-Event Models .................................... 10
           1.4.1    Petri Nets ..................................... 11
           1.4.2    Discrete-Event Specification Formalism (DEVS) ..... 11
   1.5    Experimental Frames and Model Validity ................... 12
           1.5.1    Two Capacitors Circuit ......................... 15
           1.5.2    Birth-and-death Process ........................ 16
   1.6    Model Credibility ........................................ 17
   1.7    Uncertainty and Randomness ............................... 17
   1.8    Conclusion ............................................... 18
   1.9    Questions and Answers .................................... 18
   References .................................................... 21

**2    Continuous System Models** .................................. 25
   2.1    Introduction ............................................. 25
   2.2    Dynamic Systems .......................................... 26
           2.2.1    General Classification ......................... 28
   2.3    Linearity ................................................ 29
   2.4    Ordinary Differential Equations and Models of Systems
           with Concentrated Parameters ........................... 30
   2.5    Transfer Function ........................................ 32
           2.5.1    Stability of Linear Models ..................... 34
           2.5.2    Routh–Hurwitz Stability Criterion .............. 34

2.5.3      Frequency Response ............................. 36
2.6   Nyquist Plot and Stability Criterion ........................ 38
2.7   Analog Computer Models .................................. 39
2.8   Z-transform ............................................. 40
      2.8.1      Matched Pole-zero ............................. 44
2.9   Non-linear Models and Stability ........................... 45
      2.9.1      BIBO Stability ................................. 46
      2.9.2      Lyapunov Stability ............................. 46
      2.9.3      Asymptotic Stability ........................... 46
      2.9.4      Orbital Stability ............................... 47
2.10  Stiff Equations .......................................... 51
2.11  Example: ODE Model of a Car Suspension ................. 52
2.12  Graphical Representations of Continuous Models ............. 56
      2.12.1     Block Diagrams and Signal Flow Graphs ............ 56
      2.12.2     Mason's Gain Formula .......................... 58
      2.12.3     Bond Graphs ................................... 61
      2.12.4     Example of Bond Graph ......................... 63
      2.12.5     The Causality and DYMOLA ..................... 64
2.13  Models with Distributed Parameters, Partial Differential
      Equations ............................................... 65
      2.13.1     PDE Solution Algorithms ........................ 65
      2.13.2     Finite Element Model .......................... 68
      2.13.3     Example: Jet Takeoff Vibrations ................. 69
2.14  Conclusion .............................................. 70
2.15  Questions and Answers ................................... 70
References .................................................... 78

3   Differential Inclusions, Uncertainty, and Functional Sensitivity .... 81
3.1   Introduction, Some Definitions ............................ 81
3.2   Differential Inclusions ................................... 83
3.3   Reachable Set ........................................... 83
3.4   Differential Inclusions and Control Systems ................. 85
      3.4.1      Uncertainty Treatment .......................... 86
3.5   Functional Sensitivity .................................... 87
3.6   Differential Inclusion Solver ............................. 88
      3.6.1      Example: A Second-Order Model .................. 93
3.7   Discrete Differential Inclusions ........................... 95
      3.7.1      Reachable Set, Optimal Trajectory ................ 96
      3.7.2      Example 1 ..................................... 98
      3.7.3      Example 2 ..................................... 100
3.8   Conclusion .............................................. 101
3.9   Questions and Answers ................................... 102
References .................................................... 103

**4    Functional Sensitivity Applications** ........................... 107
    4.1    Introduction ............................................. 107
    4.2    Functional Sensitivity .................................... 108
            4.2.1    Differential Inclusions .......................... 108
            4.2.2    Sensitivity Analysis ............................. 108
    4.3    Differential Inclusion Solver ............................. 110
    4.4    Example: The Lotka–Volterra Model ....................... 111
    4.5    A Mechanical System .................................... 113
    4.6    Functional Sensitivity of the V/f Speed Control of Induction
            Motor .................................................. 115
            4.6.1    Comparison with the Classical Sensitivity
                        Analysis ........................................ 119
    4.7    PID Anti-Windup Control ................................ 121
    4.8    Vehicle Horizontal Movement ............................. 127
    4.9    Marketing Sensibility and Reachable Sets .................... 130
            4.9.1    The Model ...................................... 131
            4.9.2    Experiment 1 ................................... 134
            4.9.3    Experiment 2 ................................... 135
    4.10   Conclusion .............................................. 137
    4.11   Questions and Answers ................................... 138
    References .................................................... 138

**5    Attainable Sets in Flight Control** .............................. 141
    5.1    Introduction ............................................. 141
    5.2    Control and Reachable Sets ............................... 142
            5.2.1    Airplane Dynamics ............................. 142
            5.2.2    Attainable Sets .................................. 144
    5.3    Conclusion .............................................. 147
    5.4    Questions and Answers ................................... 147
    References .................................................... 148

**6    Modeling, Simulation, and Optimization** ....................... 151
    6.1    Introduction ............................................. 151
    6.2    Landing on the Moon .................................... 153
    6.3    Iterative Algorithm ...................................... 156
    6.4    Market Optimization ..................................... 157
    6.5    Computer Implementation: Simulation and Optimization ....... 163
    6.6    Conclusion .............................................. 167
    6.7    Questions and Answers ................................... 167
    References .................................................... 169

**7    Discrete Event Models** ........................................ 171
    7.1    Introduction ............................................. 171
    7.2    The Event Queue ........................................ 173
    7.3    Agent-Based Models ..................................... 174
            7.3.1    People Agents ................................... 176

7.4    Discrete Event Specification Formalism (DEVS) .............. 177
       7.4.1    A Remark on Ambiguity .......................... 177
       7.4.2    DEVS .......................................... 178
7.5    Petri Nets ............................................... 179
7.6    Distributed Simulation Models ............................ 180
7.7    Conclusion .............................................. 182
7.8    Questions and Answers .................................. 182
References ..................................................... 185

**8    Self-Organization, Organization Dynamics, and Agent-Based
       Model** ................................................... 189
8.1    Introduction ............................................. 189
8.2    The Model .............................................. 192
       8.2.1    Interaction Rules .............................. 195
8.3    BLUESSS Simulation Package ............................ 197
8.4    Simulations ............................................. 198
8.5    Conclusion .............................................. 201
8.6    Questions and Answers .................................. 203
References ..................................................... 204

**9    The Space of Models, Semi-Discrete Events with Fuzzy Logic** ..... 207
9.1    Introduction ............................................. 207
       9.1.1    Distance Between Models ........................ 208
9.2    Strictly Discrete Event Model ............................ 209
9.3    Finite-Time Event Model ................................. 211
       9.3.1    The Chicken Game ............................. 212
       9.3.2    Semi-Discrete Model Specification ................. 214
       9.3.3    Model Coupling ................................ 217
9.4    More Examples .......................................... 218
       9.4.1    Example 1: One Server ......................... 218
       9.4.2    Example 2: Two Servers ........................ 221
       9.4.3    Example 3: A Battlefield ....................... 223
9.5    Singularity of the Exact DES Models ...................... 224
9.6    Conclusion .............................................. 226
9.7    Questions and Answers .................................. 226
References ..................................................... 227

**10   Models and Categories** .................................... 229
10.1   Introduction: The Language of Categories .................. 229
       10.1.1   Examples ...................................... 230
       10.1.2   Simultaneous Events ........................... 233
10.2   Conclusion .............................................. 234
10.3   Questions and Answers .................................. 234
References ..................................................... 235

**11 Fuzzy Time Instants and Time Model** .......................... 237
   11.1 Introduction ............................................. 237
   11.2 The Fuzzy Time Instant .................................. 238
       11.2.1 Example ....................................... 242
   11.3 Conclusion ............................................. 244
   References .................................................. 245

**12 Uncertain Future, Reversibility and the Fifth Dimension** .......... 247
   12.1 Introduction ............................................. 247
   12.2 Uncertain Future ........................................ 247
   12.3 Differential Inclusion Solver ............................. 249
   12.4 Solving the Ideal Predictor Problem. Feedback From
       the Future ............................................... 251
       12.4.1 Example 1: A Linear Model ....................... 252
       12.4.2 Example 2: A Non-Linear Model .................. 254
       12.4.3 Example 3: A Control System .................... 254
   12.5 Reversibility ............................................ 258
       12.5.1 Irreversibility of Differential Inclusions ............. 259
   12.6 Encapsulated Universe and the Fifth Dimension .............. 261
       12.6.1 General Remarks ............................... 262
       12.6.2 The Ball ....................................... 263
       12.6.3 The Metric Structure ........................... 264
       12.6.4 Linear Vector Space Operators .................... 265
       12.6.5 Local Ball and Local Observer .................... 267
       12.6.6 Velocity Superposition ......................... 268
       12.6.7 Particle Movement and a Small Bang .............. 269
       12.6.8 Adding the Time Dimension ...................... 271
       12.6.9 Uncertainty and Traveling Beyond the Infinity ....... 272
       12.6.10 The Fifth Dimension ........................... 275
       12.6.11 Conclusion .................................... 276
   References .................................................. 277

**Index** ......................................................... 279

# Chapter 1
# Concept of Model

## 1.1 Introduction: General Remarks

In this chapter, we provide a discussion about model types and related topics. A more detailed discussion about these concepts can be found in the consecutive chapters.

The use of models is a frequent and, perhaps, indispensable element of the scientific research. Recently, the growth of the amount of information available on the Internet provides a fast access to the published works on models in interdisciplinary applications. The Google search for the phrase "model of," excluding "modeling," shows more than 300 million references. The phrase "mathematical model" results in nearly 20 million positions. The number of books and articles on modeling is also very big and growing exponentially. This makes it somewhat difficult to elaborate a comprehensive survey of scientific modeling or to find the most relevant and original applications. Consequently, as stated in the Preface, we do not discuss here models being developed in some specific fields of research like models of neural networks, weather forecasting, human brain, missile dynamics, human immune system, games, electromagnetic wave propagation, human DNA, urban development, and many others. We focus on some selected and illustrative examples, and provide proposals for new tools like *differential inclusions* or *semi-discrete models*. Some original models and simulation examples are shown.

**This book treats about dynamic models**. We do not discuss static models that can be treated by algebraic methods or static modeling and optimization. This is because **the real world is dynamic**. Everything is changing, fluctuating, or oscillating. The *time* variable appears in all models we present.

One of the most important sources of modeling and simulation theory is the book of Zeigler [52] that provides basic concepts in modeling, like cellular simulation, model building, and validation. The methods of continuous system modeling are described in the book of Cellier and Greifeneder [8]. The book of Kheir [26] describes the fundamentals of modeling and simulation of continuous-time, discrete-time, discrete-event, and large-scale systems. A good collection of articles on modeling and simulation can be found in the book of Pidd [38]. These articles

© The Author(s), under exclusive license to Springer Nature Switzerland AG 2022
S. Raczynski, *Models for Research and Understanding*, Simulation Foundations, Methods and Applications, https://doi.org/10.1007/978-3-031-11926-2_1

are written by authors from academic environment who share experiences vital to readers that are seeking to expand their level of understanding in model development. Useful hints about model building and validity are proposed in the paper of Law [28].

Some useful sources on mathematical modeling and computer simulation can be found in the Springer SEMA SEAMI series. For example, Barrera et al. [4] deal with models that depend on the highly non-linear behavior of a system of partial differential equations, and adaptive reconstruction of industrial models. The interpolation methods are addressed and the *quasi-interpolation* concept is discussed. The relation between economic models and poverty problems is pointed out. In the same series, a valuable contribution is presented by Rebollo et al. [41]. In that book, we can find a collection of 30 papers from the 9th International Congress on Industrial and Applied Mathematics (Valencia, July 15–19, 2019). The papers address important topics in mathematical modeling, industrial and environmental mathematics, mathematical biology and medicine, reduced-order modeling, and cryptography.

Most of the contributions of the above book series refer to applied mathematics, in particular to distributed parameter systems and partial differential equations. Somewhat more interdisciplinary mathematical problems are discussed in "Applied Mathematics for Environmental Problems" by Ascensio et al. [2]. In the book, several models are presented, such as the wildfire spread model and wind shear forecasts.

A wide and valuable insight on construction and use of models can be found in the book of Page [36]. The author promotes the "many model thinking" approach. This consists in using ensembles of many models to understand complex phenomena. In the first chapter of the book, we read *"The logic behind the many model approach build on the age-old idea that we achieve wisdom through the multiplicity of lenses. The idea traces back to Aristotle, who wrote of the value of combining the excellencies of many."* Page explains the fundamental properties of models; a model can simplify, stripping away unnecessary details, and formalize, using mathematics. The models *"create tractable spaces within which we can work through logic, generate hypotheses, design solutions and fit data."* In the book (chapter three), Page recalls that we can apply the same model to many different cases, changing assumptions and reassigning identifiers. Indeed, if we look at the creation and the spread of the System Dynamics (SD) model of Forrester [17], we can see that the known models of electric circuits can be applied to industrial dynamics, urban growth, and a lot of other cases. This caused the growing application of the SD approach in the 1960s up to, perhaps, the recent time. However, let us observe that such approval of (any other) methodology bears some possible pitfalls. The common conviction that everything can be modeled the same way as electric circuits, and that everything that changes continuously can be described by differential equations, is not exactly true. More remarks on this issue are made further on in the present chapter.

Note that the multi-model paradigm has already been well defined in the book of Zeigler (1970) [52], where, for each basic model of a real system, there is defined a series of simplified models with their corresponding experimental frames, see the Sect. 1.5 of the present chapter.

## 1.2 The System

Modeling and simulation concepts almost always refer to the notion of *system*. By a *system*, we mean a *set of components that are interrelated and interact with each other*. These interrelations and interactions differentiate the system from its environment. The system is supposed to be organized in order to perform one or more functions or *achieve one or more specific goals*. Carter McNamara in his book [31] defines a system as follows: "*... a system is an organized collection of parts (or subsystems) that are highly integrated to accomplish an overall goal. The system has various inputs, which go through certain processes to produce certain outputs, which together, accomplish the overall desired goal for the system.*"

The commonly mentioned properties of systems include:

* Aggregation—that means that systems can be grouped into categories that can be nested into larger aggregates.

* Non-linearity—System needs not to be linear, i.e., it does not necessarily satisfy the principle of superposition.

* The behavior of a system is not the sum of the behaviors of its components.

Consult Holland [20, 21].

## 1.3 Mathematical Models

*Mathematical modeling* is one of the most important tools used in scientific research. It permits to find the mappings between the real world and the world of mathematics. In this book, we treat mathematical modeling as just one of the ways to construct models of real (or fictitious) systems. This section is about mathematical modeling. We treat a model as *non-mathematical* if it uses no mathematics, except, perhaps, elemental arithmetic, and logical relations. These are, for example, some discrete-event, object- and agent-oriented models of populations, road traffic, or battlefield, where model component actions need no high mathematics. Such models are discussed in other chapters. Note that any classification of things is, in some situations, a risky and limiting task. *Inserting all we can do to a series of boxes or drawers may limit our general insight into the real world.*

Models used in research, engineering design, forecasting, and decision-making may be of many kinds. Physical models are still used by engineers, for example, in aerodynamical problems or fluid flow simulations. Other types of models include logical, discrete-event, and object- or agent-based models used in waiting lines simulation and manufacturing of social (soft systems) problems. However, the mostly used (but not the unique) type is mathematical modeling.

According to the Encyclopædia Britannica, "*Mathematical model, either a physical representation of mathematical concepts or a mathematical representation of reality. Physical mathematical models include reproductions of plane and solid geometric figures made of cardboard, wood, plastic, or other substances; models of conic sections, curves in space, or three-dimensional surfaces of various kinds made of wire, plaster, or thread strung from frames; and models of surfaces of higher order that make it possible to visualize abstract mathematical concepts."*

We can agree or not with this definition that may appear somewhat complicated. Perhaps a more concise explanation can be found in The Free Dictionary: *"A mathematical model is a description of a system using mathematical concepts and language. The process of developing a mathematical model is termed mathematical modeling. Mathematical models are used in the natural sciences (such as physics, biology, earth science, chemistry) and engineering disciplines (such as computer science, electrical engineering), as well as in non-physical systems such as the social sciences (such as economics, psychology, sociology, political science). Mathematical models are also used in music, linguistics and philosophy (for example, intensively in analytic philosophy)."*

A model may help to explain a system, study the effects of different components, and make predictions about behavior. In the book of Eykhoff [16], we find the definition of the mathematical model as *"a representation of the essential aspects of an existing system (or a system to be constructed) which presents knowledge of that system in usable form."* However, dealing with models, we should remember that mathematical modeling is not the unique way to model building.

One of the classic sources of mathematical modeling is the handbook of G. Korn and T.M.Korn [27]. This is a big collection of useful mathematical facts related to model building. In the book, we can find definitions, theorems and formulas, and reference materials, collected with a rigorous mathematical approach. It includes the definitions and discussion of many advanced mathematical methods that may be used in modeling.

Mathematical modeling is used to solve complex problems and to understand what happens in the real world. In few words, this is a process in which real-life problems are translated into mathematical language. In this book, by *mathematical model*, we understand models that use higher mathematics, like ordinary or partial differential equations (PDE), differential inclusions, higher algebra, and advanced statistics. Models that use logical relations, arithmetics, or simple statistics used in computer simulations, will be treated rather as *computational models*. These are, for example, object-oriented queuing models or agent-based models (excluding the queuing theory).

Recently, model creation is almost always done bearing in mind computer simulation. According to the Springer Encyclopedia of Sciences of Learning, *"Computer simulation model is a computer program or algorithm which simulates changes of a modeled system in response to input signals"*. In the following, we will refer to *computer simulation* just as *simulation*. The difference between modeling and simulation tasks is that the *modeling* consists in establishing *relations between a system of real world and a model*, while *simulation* is the *relation between models and computers*. We will not discuss here the classification and tools of computer simulation because this book is focused on models, rather than simulation. Only recall that simulating something on a (digital or analog) computer is to make a computer program behave like an aeroplane, a bacteria, a factory, a moving particle, or other objects of the real (or imaginary) world. Simulation is a highly interdisciplinary research tool, where the simulationist must learn not only how to develop and use a simulation program but also how a colony of bacteria grows, how a car suspension works, how a missile moves, how shoes are fabricated, and many other things. Anyway, model building and computer simulation are closely related. In this book, we only discuss dynamic models, that describe the changes in the model state in time.

## 1.3.1   Kinds of Mathematical Models

The main, rough classification of mathematical models include

* **Linear**—this kind of model must satisfy the definition of linearity, from the mathematical point of view, namely, the *additivity and homogenity* assumption. Recall that a function $f(x)$ is additive if $f(x + y) = f(x) + f(y)$, and homogenous if $f(kx) = kf(x)$ for any $k$. Otherwise, the function is non-linear. In systems theory, big complex models are supposed to be non-linear, and the behavior of the model as a whole is not a sum of the behavior patterns of its components. The non-linearity is often associated with phenomena such as chaos and irreversibility.

* **Explicit** model—using such model, you can calculate all output variables, provided all inputs are defined. The **implicit model** requires more, frequently iterative algorithms to be used.

* **Probabilistic or stochastic**—models are those that include some randomness, represented by random variables. Otherwise, the model is **deterministic**.

* **Black box**—models are constructed looking for correct relations between model input and output, while the internal model structure is not the most important issue.

* **Complex models**—are supposed to be related with complex systems, where the model (or system) behavior can hardly be understood, and the models help us to predict the future system outcome.

* **Empirical** models—are constructed using the data obtained by observations and processed in some way, for example, by the regression analysis. The opposite kind is **mechanistic or theoretical**, where we use the relationships taken from known theories, like equations of particle movement or queuing theory.

Models can be also classified due to the particular mathematical tool:

* Models described by **ordinary differential equations (ODEs)**. This includes the System Dynamics (SD) models developed in 1960s by Jay Forrester [17] where the models are described by ODEs or difference equations. Here should also be included models given in the form of block signal diagrams, signal flow graphs, and bond graphs that automatically generate ODE versions.

* **Distributed** parameter models described by partial differential equations or finite element method.

* **Differential inclusion** models that provide reachable sets and functional sensitivity analysis (see Chap. 3).

* Models based on **regression analysis** and advanced methods of probability theory
* **Transfer functions and state transitions**.

In the automatic control theory, modeling of linear dynamic systems, electronics, instrumentation, and similar fields, the dynamics of model components are frequently described by *transfer functions*, in terms of the Laplace transform. This kind of model is discussed in more detail in Chap. 2, "Continuous models."

* Other models that **combine the above** mathematical methods or use high mathematics.

Another classification, similar to that of computer simulation, consists of the division between continuous- and discrete-event models. The point of this book is that this is somewhat artificial division that arose from the development of simulation software, rather than from the properties of the real systems being modeled. The validity of discrete-event models might be questioned, mainly because of the problems with simultaneous event handling. On the other hand, there exists a strange conviction among many researchers, that everything that is continuous in the real world, can be modeled by differential equations. The SD methodology is so user-friendly, that almost everybody can simulate everything using an SD package like Stella or PowerSim. Few users of these tools care about the model validity and the existence of solutions. This leads to invalid models that can provide wrong hints in decision-making processes.

In each field of scientific research, an intense work on modeling and simulation is going on. Each branch of research has its own approach to modeling and model classification. As an example, we give a very short overview of modeling in economy and public health.

## 1.3.2   Models of Economic Growth

The research on economic growth models dated from the eighteenth century. The early publication "An Inquiry into the Nature and Causes of the Wealth of Nations" of Adam Smith appeared in 1776, now available in other editions (Smith [45]). Smith deals with the production and distribution in a capitalistic system, being the main theoretic contribution to the pre-Marxist economy. Smith studied the formation of the capital, the productivity, and the output from workers. The factor of technical progress and its importance has been addressed later by Ricardo [43].

More mathematical models of economy dynamics appeared in the early twentieth century. Sir Roy Harrod in 1939 and Evsey Domar in 1946 developed a more detailed model of economic growth (see Sato [44]). In that model, the economic growth is dependent on the level of saving and the capital–output ratio (productivity of capital investment). The model explains how growth has occurred and how it may occur again in the future.

An interesting model that shows a constancy in the capital/output ratio, including "capital saving" and "labor saving" factors, can be found in Nicolas Kaldor [23].

The Keynesian [24] approach, developed in 1930s is used. The model also takes into account the limited available resources. Model variables are income, capital, profits, wages, investment, and savings. As the result, the model provides long-run tendencies for economic growth.

Rostow [22] proposes a five-stage development model. The process of economic growth is divided into the stages of (1) traditional society, (2) preconditions for takeoff, (3) economic takeoff, (4) drive to maturity, and (5) mass consumption. This is a long-term approach. The traditional society stage is assumed to include the eighteenth-century societies. The takeoff is considered to appear in the nineteenth century. Drive to maturity takes place in the nineteenth and twentieth centuries. Mass consumption period in US and Canada is located in 1920 and in other countries after World War II.

It is out of scope of this chapter to give a more detailed survey, consult, for example, Basu [5]. Many economic growth models use the SolowSwan model related to the Cobb–Douglas production function, proposed by Solow [46] and Swan [47], and the Bhattacharya [6] three-sector models (see also Diamand and Spencer [15]).

Walde [50] addresses the problem of optimal household behavior, including a risk factor, savings, and returns. However, the paper is focused on equilibrium rather than the extreme model behavior. Barlevy [3] discusses the design of macroeconomic policies with uncertainty. The aim is to minimize the worst-case. The uncertainty generated by the environment is taken into account. It is pointed out that the worst-case is not the most important issue, compared to the optimization of the macroeconomic policies.

The **Central Economic Mathematical Institute of URSS** was founded in 1963 to enforce the works on economic models. The main project of the institute was the "mathematical description of soviet economy." There are a few publications on these works, related to the multi-branch economy and Leontief's theory [42]. The result of the work of several years was a mathematical model that intended to describe the dynamics of the economy of the whole Soviet Imperium. However, when the model was terminated, the authors agreed that it was too big to be simulated on computers in the 1960s (machines with 32 kBytes of operational memory, 100 kHz CPUs). So, the work remains rather a theoretical achievement and did not help in slowing the problem of decay of the economy. The impact of computer simulation on the soviet economy was mentioned by several authors, see for example, the works of Naylor [32].

An example of new insight into uncertainty treatment in the economic growth models can be found in Raczynski [39].

### 1.3.3 Models in Public Health and Epidemics

One of the first classic models of the phenomenon appears in the late 1920s, in the works of Kermack and Kendrick [25]. In fact, the models that have been developed later on, are of similar type: they try to reflect the dynamics of the epidemics using

**Fig. 1.1** The SIRS model

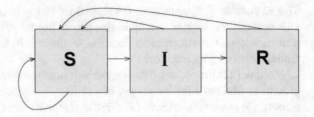

the system dynamics approach and the ordinary differential equations (ODE). The ODE models are normally given in the form of a set of non-linear ODEs and their properties. Such models have been widely discussed and improved in many available works. However, the ODE or System Dynamics (SD) models (Forrester [17]) can hardly reflect the geospatial issues. As stated in the Preface, the popularity of the SD models arose among the modelers mainly due to the strange conviction that any continuously changing variable of the real world can be modeled using the ODEs. In fact, this is not true. This, and other deficiencies of the ODE and SD modeling, have inspired the development of other modeling and simulation tools, such as the object- and agent-based simulation, see Obaidat and Papdimitrou [35], and Perez and Dragicevic [37]. Dargatz and Dragicevic [11] consider an application of an extended Susceptible–Infectious–Removed model of the "space-temporal spread of influenza in Germany. The inhomogeneous mixing of the population is taken into account by the introduction of a network of sub-regions." The multivariate diffusion process is used to describe the model.

The basic and most popular is the Susceptible–Infectious–Removed (SIR) model. Figure 1.1 shows the SD scheme of the model. Block S represents the number of susceptibles, i.e., the individuals that can be infected. Block I denotes the number of actually infected and R is the number of people who recovered from the disease with immunity, or dead. A modification of the SIR model named SIRS includes also a feedback from block R to S, as shown in Fig. 1.1.

There are many modifications to the above models. For epidemics with a larger duration, the birth-and-death process is added. A passive immunity is taken into account in the MSIR model, where it is supposed that some individuals are born with immunity. MSEIR model has the scheme MSEIR, where M is the number of passively immune individuals. Supposing that the immunity in group R is temporary, we obtain the MSEIRS model.

In the article of Ng et al. [34], we can find a description of a model of double epidemic. The two superimposed epidemics are considered using a modification of the SIR model. The problem is focused on the Hong Kong SARS (Sever Acute Respiratory Syndrome) epidemic in 2003, caused by two different viruses. The resulting model is given by a system of six differential equations of the first order.

Some more complicated models of epidemic dynamics can be found in Gebreyesus and Chang [18]. They propose a multi-compartment model that takes the interactions of human and animal or different species of animals into account. It is a multivariable state-space model that reflects the phenomena of some diseases transmitted from animal to human, such as Ebola, MERS-Coronavirus bird flu, and tuberculosis. The basic model used in that article is a modification of SIR. The main concept is to define a number of clusters in the populations, where the epidemics are governed by SIR equations, interacting with each other. The model supports the spatial issues, introducing the regions like urban and rural. For the models of recurrent epidemics, see David [13].

The above models are deterministic. In order to manage the uncertain elements, the stochastic elements in disease spread models have been introduced many years ago. Various types of stochastic models are discussed by Allen [1], who considers some modifications of the basic SIR model. The mathematical tools include the Markov chains and stochastic differential equations. From these models, we can obtain the probability distributions for final size of the S, I, and R groups, for the disease extinction, disease duration, and other parameters. In the Allen model the S, I, and R are treated as discrete random variables. A detailed mathematical background is given, where the obtained equations describe the probabilities rather than the instant values of the variables. The MATLAB code examples are given. See also Matis and Kiffe [30]. The overview of various modeling techniques, deterministic and stochastic, can be found in Daley and Gani [14]. See also Raczynski [40] for some remarks on uncertainty treatment in epidemics models.

### *1.3.4 Graphical Representations of Continuous Models*

The use of ODEs or PDEs is not the only way to simulate continuous dynamic systems. A continuous model may be constructed in the form of a graphical image, with defined parameters and initial conditions. The most used representations are *block diagrams*, *signal flow diagrams*, and *bond graphs*.

**Block diagrams** are commonly used to describe automatic control system and instrumentation problems.

**Signal flow diagrams** define the signal flow in the model, using arrows and transfer function. Static linear and non-linear arrows are also used. See Tavangarian and Waldschmidt [48].

**Bond graphs** are used to simulate the dynamics of physical systems. In this method, we use links in form of directed *bonds* like harpoons. The bond is associated with a corresponding *flow* (for example, velocity or electric current) and *effort* (for example, the force or electric voltage). Frequently, a bond represents the flow of energy in the real system. See Cellier [7].

The advantage of the above methods is that there are software tools that automatically generate the model ODEs, and then, the corresponding computer code. See Chap. 2 for a more detailed description of these types of models. Models of systems with distributed parameters described by partial differential equations or the finite-element method are also discussed in Chap. 2.

We should not restrict our modeling methodology to the known tools, as discussed above. One of the alternative and not commonly modeling tools is the **differential inclusion**. Models of such kind are discussed in Chap. 3.

### 1.3.5   Computational Tractability

While creating a model, we should clearly define the purpose of our task. A model can be created to be used in computer simulation or for other possible use, for example, stability analysis, a mathematical representation of a real system, or just as an abstract model that may explain some general ideas. For models that will be used in consecutive simulation tasks, the problem of *computational tractability* arises.

It is well known in computer simulation and operations research that some problems that have a nice mathematical description cannot be treated using the hardware we actually have. This is very important for models prepared to be run on a computer as a simulation task. Some models are computationally intractable, which means that they are too time-consuming or expensive to be analyzed on the computer.

A modeling and simulation task is *intractable* if its computational complexity increases exponentially with the number of its descriptive variables. The *computational complexity* can roughly be defined as the minimal cost of guaranteeing that the computed answer to our question (simulation task) is within a required error threshold.

A mostly cited example of an intractable problem is the salesman problem, namely the simulation of all possible routes of a salesman that must visit various cities. The aim is to find the route with the shortest total distance. Other examples of continuous system simulation can be found in fluids dynamics applied to problems like the reentry of the space shuttle to the atmosphere of the earth (modeling the airflow around the craft).

For more discussion on intractability, consult Traub and Wozniakowski [49] or Werschulz [51].

### 1.4   Discrete-Event Models

Other models, different from the mathematical models discussed above, are the *discrete-event models*. In models of such kind, the *model time* (time to be modeled, not the time of our clocks) jumps from one *model event* to another and does not advance continuously or in small time-steps. This makes the corresponding computer simulation fast, compared to the ODE model simulation. A more detailed discussion of discrete-event model can be found in Chap. 7.

The discrete-event models are given in the form of the specification of possible events that may occur in the modeled system, and in managing the event execution. The order of execution is not predefined, and is defined at the runtime, by the corresponding software package, and not by the user. So, each simulation tool, like GPSS or Arena package, is, in fact, a certain version of discrete-event model, ready to use for a class of similar modeling tasks, like queuing models.

Discrete models are closely related to discrete-event simulation. Creating a discrete model, we almost always do it using a particular discrete-event simulation software. As stated before, modeling and simulation are closely related to each other. For mathematical models, we have perhaps more freedom in simulation while selecting a simulation software, like a numerical method or continuous simulation package. For discrete models, we almost always select the software tool before constructing the model. Of course, a discrete-event execution can be coded in an algorithmic or object-oriented language like C++. However, using, for example, the antique package GPSS, we need 20 (or more) times less code lines than in the code created (from the very beginning) in C++.

As mentioned before, the creation of models and computer simulation are sometimes inseparable tasks. From this point of view, each one of the simulation packages like GPSS [19], Arena, or ProModel represents its own modeling methodology. In discrete-event simulation, the very elemental concept is that of object-oriented modeling (OOM) and related programming methods. The idea of OOM and model event execution dates from the 1960s and was also defined in the GPSS (General Purpose Simulation System) package. The GPSS objects, named *transactions*, are created at the runtime. They pass through the model events (GPSS resources), interact with each other, and disappear. This concept repeats in nearly all discrete-event simulation tools. See Chap. 7 for more detail.

### 1.4.1  Petri Nets

Petri nets (PNs) is a graphical modeling tool for discrete-event simulation. A good review of the method can be found in David and Alla [13]. PNs were originally developed by Carl Petri in 1962 to model and analyze communication systems. In PNs, there are four elements: places (represented by circles), transitions (represented by bars), directed acrs, and tokens (represented by dots). More details on Petri nets are provided in Chap. 7.

### 1.4.2  Discrete-Event Specification Formalism (DEVS)

The theoretical base of discrete-event simulation is defined in the Discrete-event Specification Formalism (DEVS), see Chow [9].

In the *discrete-event simulation*, it is supposed that the model events are discrete, i.e., they are accomplished within a model time interval of length zero. This model simplification makes the simulations very fast. In the *object-oriented programming*, we declare several generic code segments called classes. According to these declarations, objects are created at the runtime. Each object is equipped with a data set and several methods that perform operations on the data.

The *Discrete-event Specification* (DEVS) formalism is used to describe models in discrete-event simulation. In the DEVS formalism, an *atomic* model is defined, (Zeigler [52]) and then, the *coupled* model that integrates several atomic and other coupled models. This results in a hierarchical model building that uses portability and model "encapsulation," useful while dealing with big, complex systems.

In DEVS, as well as in any other discrete-event models, the problem of possible *simultaneous events* arises. To solve it, an additional model element *select* is added. The select component defines the order of execution for simultaneous events that may occur in the coupled model. This component must be added to the model to avoid ambiguities in the simulation algorithm and to make the model implementation-independent. There is a big research done on the select algorithms because treating the simultaneous events is rather a difficult task.

The use of the DEVS formalism is relevant for big models, where the time of execution, hierarchical model building, and portability are important factors. To treat complex models with variable structure, the *Dynamic Structure Discrete-event System Specification* (DSDEVS) is used.

See Chap. 7 for other remarks on discrete models.

## 1.5   Experimental Frames and Model Validity

First of all, we must define the *components* as elemental parts of the model, like clients in a bank, ships entering a harbor, cars on a street, etc. Each component is described by the *descriptive variables* that include input, component state, and output variables. The set of all descriptive variables in a model forms the *experimental frame*. The same real system may have several different experimental frames. Each experimental frame results in the corresponding simplified model, as shown on the Fig. 1.2 (following Zeigler [52]).

Here, the basic model reflects exactly the behavior of the real system. Such model normally does not exist.

Note that the aim of the modeling task, as well as the technical limitations (the computer on which we want to run the resulting simulation program), reduce the number of possible simplified models. This helps us selecting the appropriate simplification. If there is more than one simplification that satisfies our requirements, we must apply other selection rules (e.g., modeling cost). If no model exists satisfying our aim and technical limitation, then no simulation is possible. Remember that looking for something that does not exist is simply a waste of time. Also, note that the same real (or imaginary) system can have several different experimental frames and several simplified models. For example, while modeling an electric circuit, the common experimental frame is the set of the voltages and currents on the corresponding circuit components. However, for the same circuit, someone can define the experimental frame as the set of all voltages and currents, power dissipated on each

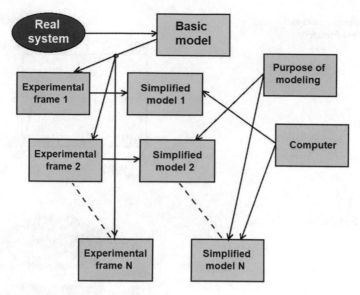

**Fig. 1.2** Basic model and experimental frames

element, the temperature of each integrated circuit and of the printed circuit plate, as well as the intensity of the electromagnetic field produced by the circuit. The first frame suggests the use of an appropriate package for circuit simulation, while the second one implicates the use of a circuit simulation package, as well as sophisticated heat transfer and electromagnetic wave software.

Model *validity* is one of the central problems in modeling and simulation. Before discussing the concept, recall what the input and output variables are. Roughly speaking, by the *input* we mean all external signals like electric excitations, control signals, or orders that come from the environment. *Output variables* are those values that we want to observe, measure, store, print, or plot as the result of a simulation run. The concept of input and output comes from the problems of signal processing, automatic control, and similar fields. However, not all systems must be causal, and the input and output concept may not work. Consider an electric resistor, treated as an isolated, stand-alone system. One can define the input signal as the voltage applied to the resistor and the output as the resulting current. But the same resistor in other circuits (context) can have a forced current (connected to a current source) as the input signal, the resulting output variable being the voltage. A new approach to modeling of physical systems questions the input–output concept, which means that from this point of view, physical systems are not causal (see Cellier [8]). Other aspects of model descriptive variables and formal model definition will be discussed in the section devoted to the DEVS (Discrete-event Specification) formalism.

Now, let us come back to causal systems with input and output signals well defined. Consider a real dynamic system and a corresponding model. Let $S$ be the operation of

**Fig. 1.3** Basic model and
experimental frames

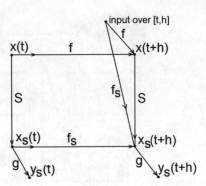

**Fig. 1.4** Input/output
validation

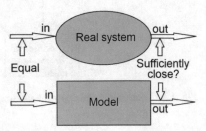

modeling (passing from real system to its model). By $x(t)$ we will denote the system
state in the time instant $t$, by $y(t)$ the system output, by $f$ the state transition function
that maps the state $x(t)$ and input over $[t, t+h]$ into the new state $x(t+h)$. The same
letters with suffix $s$ denote the corresponding items for the model. Due to Zeigler
[52], the model is said to be *valid* if and only if the diagram of Fig. 1.3 commutes.
In other words, starting with $x(t)$, we must obtain the same model output $y_s(t+h)$
independently of the way we choose. This must be satisfied for any possible initial
state and input. This definition of model validity is somewhat difficult for practical
applications. A more practical concept is the *input–output or external validity* that
can be illustrated by the scheme shown on Fig. 1.4.

The model is supposed to be I/O valid if the outputs from the model and from
the real system are "sufficiently" near. What "sufficiently" means is the individual
judgement of the modeler. Obviously, the above property must be satisfied for a long
time interval, and perhaps for future model inputs. As we do not know what inputs
signals will affect our model, the only way to check the I/O validity is to compare
the model behavior and the real system output using some historic data.

The validation methods are discussed in a huge number of publications. Let us
cite only one of them that provides a quite recent and comprehensive survey, see
Ling and Mahadevan [29].

**Fig. 1.5** Two capacitances

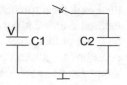

## 1.5.1 Two Capacitors Circuit

Consider an electric circuit of Fig. 1.5 and two experimental frames:

Frame 1. Variables: voltages on the two capacitances.

Frame 2. Variables: voltages on the two capacitances and the total energy in the circuit.

In this model $C_1 = C_2$, the wires have a resistance equal to zero, and the circuit has no inductance.

Suppose that at the beginning, the voltage of $C_1$ is equal to V, and the $C_2$ has voltage zero (Fig. 1.5). At a certain time instant, we close the switch. Logically, after the switch is closed, the electric charge distributes between the two capacitances so that each of them has the voltage equal to V/2. Obviously, the model is valid for experimental frame 1.

Now, calculate the total energy before and after closing the switch. The result of this simple calculation is that half of the energy disappears. The question is, where the half of the energy went? This means that the model is invalid for experimental frame number 2.

To build a valid model, we must consider a (may be arbitrary small) wire resistance $R$, or a finite circuit inductance. If $R$ is greater than zero, then half of the energy dissipates on $R$. It is easy to see that this amount of energy **does not depend on** $R$. Now, suppose that $R$ tends to zero. One could expect that, in the limit ($R$=0), the dissipated energy is the same, but this is not the case because for $R$= 0 no energy dissipates. In other words, the sequence of models with $R$ approaching zero tends to a singularity at $R$=0. This, and similar issues, are discussed in Chap. 9.

Unfortunately, for more complex models, we can never be sure that our model is 100% valid. The exact validity criterion of Zegler is difficult to implement. The I/O validity only shows that the model is good for the past data. We can "extrapolate" this assessment with some probability to the future model behavior. However, in the future, the data may change abruptly or the real system may change its structure and interaction rules so that simulation results will be wrong.

As an example, consider a model of a new epidemy, like COVID-19. I appreciate efforts on COVID-19 pandemics simulations that perhaps can result in some useful tools in the future. However, note that the most *serious mistake* in any scientific work is *to look for something that does not exist*. The problem of the existence of solutions is well known in mathematics but somewhat vague in other fields of scientific research. For new epidemics with new forms of bacteria or viruses, the existence of the model is rather doubtful. The worst error is to take a plot of certain past epidemics and then look for a forecast by using the data of the initial period and use some best-fit method to estimate parameters and generate forecasts. Such curves and forecasts may be (and, in fact, they were) used by health organizations and governments to take decisions on future actions. This may result in *erroneous disease handling* and in thousands of infections and deaths. Anyway, this is a question of ethics in modeling and simulation.

Note that the same mathematical model may describe two or more real systems. For example, the equation of an electrical capacitance can describe, as well, the dynamics of a cylindrical water deposit. The analogy between electrical circuits and the real manufacturing, business, social, or economical systems was used by Jay Forrester [17] in the development of the System Dynamics methodology.

On the other hand, a real system may be described by two or more models of a completely different kind. Let's see the following models of the *birth-and-death* (B-D) process.

### 1.5.2   Birth-and-death Process

The birth-and-death (B-D) equation may describe the behavior of growing population, queuing systems, communication systems, and similar models in a probabilistic way. The solution to the B-D equation is given by a set of functions of time $P_n(t), n = 1, 2, 3, ...N$, where $P_n(t)$ is the probability that the system takes the n-th state at the time instant $t$. The total number $N$ of possible states is, in most practical cases finite, but in theory it can be infinite. Note the difference between the experimental frame of the B-D equation model and that of a model based on discrete events. While modeling, for example, a growth of a population we can create a discrete model (A), where each event of birth and death of an entity is simulated at a discrete (deterministic or random) time instant, or describe it by a corresponding B-D equation (B-D model). The result of the model (A) is a sequence of random events, different in each simulation run. In turn, the result of the B-D equation model is a sequence of continuous functions $P_n(t)$ that describe model state probabilities rather than particular states. Note that the two corresponding experimental frames are complete different from each other and even do not intersect, one being stochastic and discrete, and the other is deterministic, represented by a set of ordinary differential equations and a sequence of continuous functions of time.

To obtain the B-D equation of given system, we must define the probability of "birth" (birth of a man, appearance of call request in a communication system, appearance of a new client in a bank, etc.), and the probability of the "death" or disappearance of a model entity. Then, an assumption is made that the probability that two events occur within a time interval H tends to zero more quickly than the probability that only one event will occur within H, when H approaches zero. This leads to the equation 1.1 (the case of population growth):

$$\begin{cases} \dfrac{dP_n}{dt} = -(\lambda_n + \mu_n)P_n(t) + \lambda_{n-1}P_{n-1}(t) + \mu_{n+1}P_{n+1}(t) \ for \ n > 0 \\ \\ \dfrac{dP_0}{dt} = -\lambda_0 + \mu_1 P_1(t), \end{cases} \tag{1.1}$$

where $\lambda_n$ is the birth rate and $\mu_n$ is the death rate if the model state is n (in this case the total population). $P_n(t)$ is the probability that the size of the population is equal to $n$ at time instant $t$.

Both the birth-and-death rates can depend on the state $n$ and on the time. For example, in the population model, the birth rate may be proportional to the number of population members, and in a communication system model the birth rate is constant or depends on external factors (the number of requests does not depend on how many clients are being attended by the system at the moment).

The B-D equation is a useful tool in modeling and simulation. However, in many practical situations it is too primitive to describe the system behavior. Normally, to derive the B-D equation one must first find how the birth-and-death rates depend on the system state. This step is frequently the source of errors. For example, while modeling a biological or ecological system the modeler frequently forgets that such systems have a huge and complicated memory. Making simple assumptions on the birth-and-death rates discards this memory and leads to serious errors.

## 1.6 Model Credibility

The end-user of a model need not be the same person or team that created the model. If so, the problem of *model credibility* should be addressed. If the end-user or client (a plant engineer, a manager, a decision-maker) who received the model and/or final simulation program, must believe that it is valid and works satisfactory. If not, the **model will never be used**. After developing the simulation model, we must exhaustively test it using different data sets. Even if we do this, we have no guarantee that the model will work. It is highly probable that the first thing the end-user will do, is to run the program with input data that makes the program crash or provides unacceptable results.

One of the things we can do to increase model credibility, is to include end-user person(s) in the team that elaborates the model.

## 1.7 Uncertainty and Randomness

In the environment of systems we want to model and simulate, as well in the system itself, there is a lot of factors that influence model behavior, but are not exactly defined. It is a common practice to treat such factors as random variables. This results in *stochastic models* that include random variables. Recall that any *random variable* should obey some rules of the theory of probability. A random value must be equipped with corresponding probabilistic properties, like the expected value, variance, probability distribution, and others. *Stochastic models* include such random variables. This means that the resulting model trajectories are also charged with certain randomness. The numerical treatment of such type of models is more difficult that the integration of deterministic models, see Nelson [33].

However, stochastic modeling is not the only way to treat system with non-deterministic components. In Chap. 3, we can find the problem statement of uncertainty caused by variables that are not random or stochastic. Such variables are just uncertain and have no probabilistic properties. Uncertain variables need not be random. It may be, for example, an external disturbance that changes abruptly in certain unknown time instant, or a value inserted intentionally into the model (false information) by external agents.

In Chap. 3, we propose an application of *differential inclusions* in uncertainty treatment.

## 1.8 Conclusion

Modeling and simulation is a highly interdisciplinary field of research. Most of the models we create are then used in computer simulation. The diversity of models is the consequence of the diversity of the real world. Moreover, as mentioned in this chapter, the same real system may correspond several, quite different models. This depends on the aim of the modeling and on the experimental frame we use.

Creating a model we must first define the purpose of the whole task. This will determine one or more experimental frames. In general, model creation helps us in understanding how complex systems work. Not all models we create result in practical applications like computer simulation, optimization, predictions, or decision-making. There are models that cannot be implemented because of the computational tractability, or are created for other purposes. Such models, however, are not completely useless. They may have cognitive properties, stimulate our imagination, or simply let us analyze and understand the real world.

## 1.9 Questions and Answers

**Question 1.1** What does it mean that a function $y(x)$ is linear?

**Question 1.2** Is the function $y = ax + b, \; b \neq 0$ linear?

**Question 1.3** Which of the following differential equations may represent a linear model of a dynamic system? ($u(t)$ is the external excitation)

$$1. \quad \frac{d^2x}{dt^2} + 4\frac{dx}{dt}(1 + x(t)) = u(t)$$

$$2. \quad \frac{d^2x}{dt^2} + 2\frac{dx}{dt} + x(t) = u(t)^2$$

$$3. \quad \frac{d^2x}{dt^2} + 5\left(\frac{dx}{dt}\right)^3 + x(t) = u(t)$$

$$4. \quad \frac{d^2x}{dt^2} + 4\frac{d^3x}{dt^3} + t^2 + x(t) = u(t)$$

**Question 1.4**  What do we understand by a *system*?

**Question 1.5**  What does it mean that a system is complex?

**Question 1.6**  What is DEVS?

**Question 1.7**  What is the *select* model element in DEVS?

**Question 1.8**  What is *experimental frame*?

**Question 1.9**  Can a valid model be not configurable?

**Question 1.10**  What is validity I/O?

**Question 1.11**  Consider two model simplifications, where simplification B is more detailed and advanced than simplification A. Must the experimental frame of A be included in the frame of B?

**Question 1.12**  What is the main field of applications of the *signal flow graphs?*

**Question 1.13**  What is the main field of applications of the *bond graphs*?

**Question 1.14**  What is a "computationally intractable" task?

**Question 1.15**  What is the *model time*?

**Question 1.16**  What is the *discrete-event simulation*?

**Question 1.17**  What are Petri Nets?

**Question 1.18**  What is the *uncertainty* in modeling a simulation?

**Answers**

**Answer 1.1**  To be linear, the function $f(x)$ must be additive and homogenous. This occurs if $f(x + y) = f(x) + f(y)$, and $f(kx) = kf(x)$ for any $k$.

**Answer 1.2** No, this is an affine function that does not satisfy the conditions of additivity and homogeneity. However, a simple change of the origin of the coordinates system that makes $b = 0$, converts it into a linear function.

**Answer 1.3** Options "2" and "4" because in these equations the independent variable and its derivatives appear in linear expressions.

**Answer 1.4** By a *system*, we mean a *set of components that are interrelated and interact with each other*. These interrelations and interactions differentiate the system from its environment. The system is supposed to be organized in order to perform one or more functions or *achieve one or more specific goals*.

**Answer 1.5** A system is considered *complex* if its behavior can hardly be understood. Models of complex systems help us to predict the future system outcome and better understand its behavior.

**Answer 1.6** The Discrete-event Specification (DEVS) formalism is used to describe models in discrete-event simulation. In the DEVS formalism, an *atomic* model is defined, and then, the *coupled* model integrates several atomic and other coupled models. This results in hierarchical model building, that uses portability and model "encapsulation" useful while dealing with big, complex systems.

**Answer 1.7** The *select* element is necessary to avoid ambiguity that may occur while modeling simultaneous events.

**Answer 1.8** The *experimental frame* is the set of all descriptive variables of the model. Roughly speaking, the experimental frame defines what we want to see, measure, and take into account in a particular model. The same real system may have several different experimental frames. Each experimental frame results in the corresponding simplified model.

**Answer 1.9** Yes.

The validity is a property of the model, while confiability refers to the relation between the model and the end-user. These are two quite different concepts.

**Answer 1.10** The model is supposed I/O valid if the outputs from the model and from the real system, taken from the past data, are "sufficiently" near.

**Answer 1.11** No. Two experimental frames of the same real system may be completely different. For example, the birth-and-death process can be simulated as a sequence of events (discrete-event simulation), or as a continuous, differential equation model that describes the changes in the probability of reaching a specific number of members of the population.

**Answer 1.12** Models of automatic control circuits, instrumentation.

**Answer 1.13** Models of physical systems with flows, efforts, and energy exchange.

**Answer 1.14** A modeling and simulation task is *intractable* if its computational complexity increases exponentially with the number of descriptive variables. A mostly cited example of an intractable problem is the salesman problem, namely, the simulation of all possible routes of a salesman that must visit various cities.

**Answer 1.15** It is the *time to be modeled*, not the time of our or computer clock In most simulation packages, the model time is represented by a variable, which value increases during the simulation.

**Answer 1.16** In models of such kind, the *model time* (time to be modeled, not the time of our clocks) jumps from one *model event* to another and does not advance continuously or in small time-steps.

**Answer 1.17** Petri Nets is a graphical modeling tool for discrete-event simulation. In Petri Nets, there are four elements: places (represented by circles), transitions (represented by bars), directed acrs, and tokens (represented by dots). The discrete events occur when the corresponding transition is activated.

**Answer 1.18** Models may include parameters that have *uncertain* values. The common approach to uncertainty is to treat these parameters as random variables. However, the uncertain parameter is not the same as a random parameter. Uncertain values may have no average, variance, or other probabilistic characteristics. Such parameters or descriptive variables are known as *tychastic* variables.

# References

1. Allen LJS, Allen EJ (1945) An introduction to stochastic epidemic models. Springer-Verlag, Berlin Heidelberg
2. Asencio MI, Oliver A, Serrate J (2019) Applied mathematics for environmental problems. Springer SEMA SEMAI series vol 6. ISBN: 978-3-030-61794-3
3. Barlevy G, Fisher JDM (2010) Mortgage choices and housing speculation. Working Paper Series WP-2010-12, Federal Reserve Bank of Chicago
4. Barrera D, Remogna S, Shibib D (2022) Mathematical and computational methods for modeling, approximation and simulation. SEMA SEMAI Springer Series. ISSN 2199-3041
5. Basu D (2009) Economic models: methods, theory and applications. World Scientific
6. Bhattacharya PC (2007) Informal sector. Income inequality and economic development, centre for economic reform and transformation
7. Cellier FE (1992) Hierarchical non-linear bond graphs: a unified methodology for modeling complex physical systems. Simulation 55(4):230–248. https://doi.org/10.1177/003754979205800404
8. Cellier FE, Greifeneder J (1991) Continuous system modeling. Springer. ISBN/ISSN 978-1-4757-3922-0
9. Chow AC (1996) A parallel, hierarchical, modular modeling formalism and its distributed simulator. Trans Soc Comput Simul 13(2):55–67. The Society for Modeling and Simulation
10. Dahl O, Nygaard B (1967) Simula–an Algol-based simulation language. Commun ACM 9:671–678
11. Dargatz C, Dragicevic S (2006) A diffusion approximation for an epidemic model. Collaborative Research Center 386

12. David JD (1945) A light introduction to modeling recurrent epidemics. In: Mathematical epidemiology Series Lecture Notes in Mathematics
13. David R, Alla H (1945) Petri nets for modeling of dynamic systems: a survey. Automatica 30(2), Elsevier. https://doi.org/10.1016/0005-1098(94)90024-8
14. Daley DJ, Gani J (2001) Epidemic modeling: an introduction. Cambridge University Press. ISBN: 978-0-521-01467-0
15. Diamand RW, Spencer BJ (2008) Trevor swan and the neoclassical growth model. Report NBER Papers in Economic Fluctuations and Growth
16. Eykhoff P (1974) System identification: Parameter and state estimation. Wiley-Interscience, London. ISSN 0-471-24980-7
17. Forrester JW (1961) Industrial dynamics. Pegasus Communications, Waltham, MA
18. Gebreyesus KD, Chang CH (2015) Infectious diseases dynamics and complexity: multicompartment and multivariate state-space modeling. In: Proceedings of the world congress on engineering and computer science
19. Gordon G (1975) The application of GPSS to discrete system simulation. Prentice-Hall
20. Holland JH (1995) Hidden order. Addison-Wesley
21. Holland JH (1998) Emergence: from chaos to order. Addison-Wesley Publishing Company, Helix Books
22. Kaya VP (2004) Rostow's stages of development. Internet communication. http://www.nvcc.edu/home/nvfordc/econdev/introduction/stages.html
23. Kaldor N (1957) A model of economic growth. Econ J 67(268):591–624
24. Keynes M (1964) The general theory of employment. Interest and Money, Harcourt Brace and Company, New York
25. Kermack WO, McKendrick AG (1927) Contributions to the mathematical theory of epidemics, part I. In: Proceedings of the royal society of edinburgh. Section A. Mathematics, the Royal Society of Edinburgh, Edinburgh
26. Kheir N (1995) Systems modeling and computer simulation. CRC Press, ISBN/ISSN, p 9780824794217
27. Korn G, Korn TM (1968) Mathematical handbook for scientists and engineers. McGraw-Hill Book Co, New York
28. Law AM (2009) How to build valid and credible simulation models. In Rossetti MD, Hill RR, Johansson B, Dunkin A, Ingalls RG (eds) Proceedings of the 2009 winter simulation conference
29. Ling Y, Mahadevan S (2013) Quantitative model validation techniques: New insights. Reliab Eng Syst Saf 111:217–231. https://doi.org/10.1016/j.ress.2012.11.011
30. Matis JH, Kiffe TR (2000) Stochastic population models. A compartmental perspective, Springer
31. McNamara C (2006) Field guide to consulting and organizational development. Authenticity Consulting, LLC. 10 1933719206
32. Naylor TH (1988) The impact of simulation on soviet economic reforms. Simulation 51(2):46–51
33. Nelson BI (2010) Stochastic modeling: analysis and simulation. Dover Books on Mathematics, Dover Publications, New York. ISSN 10.0-486-47770-3
34. Ng TW, Turinici G, Danchin A (2003) A double epidemic model for the SARS propagation. J Neg Res BioMed Cent 3(19). https://doi.org/10.1186/1471-2334-3-19
35. Obaidat MS, Papadimitriou GI (2003) Applied system simulation methodologies and applications. Springer. ISBN: 978-1-4613-4843-6
36. Page SE (2018) The model thinker. Hachette Book Group, New York. 978-0-465-009462-2
37. Perez L, Dragicevic S (2009) An agent-based approach for modeling dynamics of contagious disease spread. BioMed Cent 8(50). https://doi.org/10.1186/1476-072X-8-50
38. Pidd M (2004) Systems modeling: theory and practice. Wiley. ISBN/ISSN 978-0-470-86731-0
39. Raczynski S (2019) Dynamics of economic growth: uncertainty treatment using differential inclusions. MethodsX 6:615–632. https://doi.org/10.1016/j.mex.2019.02.029
40. Raczynski S (2020) Uncertainty in public health models treated by differential inclusions. Int J Model Simul Sci Comput 11(4). https://doi.org/10.1142/S1793962320500403

41. Rebollo TC, Donal R, Higueras I (2022) Recent advances in industrial and applied mathematics. SEMA SIMAI Springer Series. ISSN 2199-3041
42. Reyes FA (2016) Equilibrium and economic policy in W. Leontief's theory. Cahiers d'economie Politique 71:203–217. https://doi.org/10.3917/cep.071.0203. https://www.cairn.info/revue-cahiers-d-economie-politique-1-2016-2-page-203.htm
43. Ricardo D (1996) Principles of political economy and taxation. Amherst, NY
44. Sato R (1964) Harrod domar growth model. Econ J Wiley Royal Econ Soc 74(294):380–387. http://www.jstor.org/stable/2228485, https://doi.org/10.2307/2228485
45. Smith A (1998) An inquiry into the nature and causes of the wealth of nations. In: Campbell RH, Skinnes AS (eds.) Glasgow
46. Solow RM (1956) A Contribution to the theory of economic growth. Econ 70(1):65–94. http://www.jstor.org/stable/1884513
47. Swan TW (1956) Economic growth and capital accumulation. Econ Rec 32(63):334–361
48. Tavangarian D, Waldschmidt K (1980) Signal flow graphs for network simulation. Simulation 34(3):79–92. https://doi.org/10.1177/003754978003400308
49. Traub JF, Wozniakowski H (1994) Braking intractability. Scientific American, January 1994
50. Walde K (1999) A model of creative destruction with undiversifiable risk and optimising households. Econ J 109(454):156–171. https://doi.org/10.1111/1468-0297.00423
51. Werschulz AG (1991) The computational complexity of differential and integral equations: an information-based approach. Oxford University Press
52. Zeigler BP (1976) Theory of modeling and simulation. Wiley-Interscience, New York

# Chapter 2
# Continuous System Models

## 2.1 Introduction

In this chapter, some general concepts on continuous models are discussed. This chapter refers both to models and to computer simulation. Classification of dynamic systems is reviewed and summarized. The main numerical methods for the concentrated parameter systems, governed by the ordinary differential equations are described. An example of a simulation task of a simple mechanical system is given. The methods of signal flow graphs and bond graphs are discussed. A new, alternate approach is proposed, using the differential inclusions instead of ordinary differential equations. Next, there are some remarks on the distributed parameter systems, partial differential equation models, and the finite element method. We do not discuss here software tools for continuous models. Any software described in publications of such type may result to be obsolete within a few years, while the more general concepts do not change so quickly.

Continuous models include those of *concentrated parameters* and *distributed parameters* systems. The former group of models represents models for which the power of the set of all possible states (or, more precisely, the number of classes of equivalence of inputs, see Chap. 1) is equal to the power of the set of real numbers, and the latter refers to systems for which this set is greater than the set of reals. These systems will be described in more detail in Sect. 2.2. The most common mathematical tools for continuous modeling and simulation are the ordinary differential equations (ODEs) and the partial differential equations (PDEs).

As for the implementation of continuous models in computer simulation, we must remember that in the digital computer nothing is continuous, so the continuous simulation using this hardware is an illusion. Historically, the first (and only) devices that could simulate continuous models were the *analog computers*. These machines are able to simulate truly continuous and parallel processes. The development of digital machines made it necessary to look for new numerical methods and their implementations. This aim has been achieved to some extent, so we have quite good software tools for continuous model simulation.

S. Raczynski, *Models for Research and Understanding*, Simulation Foundations, Methods and Applications, https://doi.org/10.1007/978-3-031-11926-2_2

To illustrate the very elemental reason why the continuous model simulation on a digital computer is only an approximation of the real system dynamics, consider a simple model of an integrator. This is a continuous device that receives an input signal and provides the output as the integral of the input. The differential equation that describes the device is

$$\frac{dx}{dt} = u(t),$$
(2.1)

where $u$ is the input signal and $x$ is the output. The obvious and the most simple algorithm that can be applied on a digital computer is to discretize the time variable and advance the time from zero to the desired final time in small intervals $h$. The iterative formula can be

$$x(t + h) = x(t) + hu(t)$$
(2.2)

given the initial condition $x(0)=0$.

This is a simple *rectangle rule* that approximates the area below the curve $u(t)$ using a series of rectangles. The result is always charged with a certain error. From the mathematical point of view, this algorithm is quite good for input signals regular enough. Theoretically, if $u(t) \equiv 1$, the error tends to zero when $h$ approaches zero, so we can obtain any required accuracy. For other input functions, the errors may accumulate. However, even for the unit step function $u$, this does not work.

Suppose that our task is to simulate the integrator over the time interval $[0, 1]$ with $u = const = 1$. We want to implement the above algorithm on a computer on which the real numbers are stored with the resolution of eight significant digits. To achieve high accuracy of the simulation, we execute the corresponding program of formula (2.2) several times, with $h$ approaching zero. One can expect that the error will also approach zero. Unfortunately, this is not the case. Observe that, if $h < 0.000000001$ the result of the sum operation at the right-hand side of (2.2) is equal to $x(t)$ instead of $x(t) + hu(t)$ because of the arithmetic resolution of the computer. So, the error does not tend to zero when h becomes small, and the final result maybe zero instead of one (integral of 1 over $[0, 1]$). Of course, we have a huge number of numerical methods that guarantee sufficiently small errors and are used with good results. Anyway, we must be careful with any numerical algorithm and be aware of the requirements regarding the simulated signals to avoid serious methodological errors. A simple fact that we always must take into account is that, **in a digital computer, real numbers do not exist, and they are always represented as their rough approximations**.

## 2.2  Dynamic Systems

A *system* is a collection of elements or components that are organized for a common purpose.

Carter McNamara in his book [25] gives somewhat larger definition: "... *a system is an organized collection of parts (or subsystems) that are highly integrated to accomplish an overall goal. The system has various inputs, which go through certain processes to produce certain outputs, which together, accomplish the overall desired goal for the system.*"

The aim of modeling and simulation is to observe the changes in the model state over a given time interval. The fundamental concepts of the models of *dynamic systems* are the model state and causality. Roughly speaking, the *system state* is the minimal set of data that permits to calculate the future system trajectory, given the actual state and all system inputs (external excitations) over the time interval under consideration. A dynamic system may be considered causal if its actual state depends on previous states and the previous and actual external excitations only. Another notion of causality is the input–output causality relation used in signal processing, instrumentation, and automatic control. Note that these two causality concepts are quite different. An electronic amplifier has its input and output signals well defined and, obviously, the input is the cause, the output is the result, and not vice versa. However, in physical systems, this concept does not work. For an electric resistor mentioned before, the relevant variables are the current and the applied voltage. However, we cannot say which of these variables is the cause and which one is the result. If you do not define to what device the resistor is connected (e.g., voltage source or current source), you cannot say if the cause is the current or the voltage.

Consider a dynamic system with input and output signals for this system as $U$ and $Y$, respectively. We define by $X(t)$ the state of the system at the moment $t$ if and only if a function or functional $F$ exists such that

$$X(t) = F(t, U\{s, t\}, s, X(s)), \quad s < t.$$

In other words, it is necessary that the system state at a moment $t$ can be calculated using some past state in time instant $s$ and the input function over the interval between the two moments of time. For example, the state of a system that has one spring, one mass, and one damper linked together is given by the mass position and velocity. All the other descriptive variables of this system are parameters, inputs (e.g., external forces), or some output functions defined by the modeler.

In the case of an electric red composed of any number of resistors and one capacitor, the state is the value of the capacitor voltage. In this case, all the currents in the resistors can be calculated provided we know the initial capacitor voltage and the external excitations (input signals). System state may be a scalar, a vector, or an element of a certain more abstract space. For example, the state of the model describing the changes in the temperature distribution inside a piece of metal belongs to the space of all differentiable functions of three variables defined in the region occupied by the modeled body.

Now, let us define what we mean by the phrase "equivalent to," applied to functions. Obviously, two functions equal to each other in all points of a given interval, are equivalent to each other. In general, we treat two input signals as *equivalent*, if they produce the same output. For example, if the system is an integrator, two input signals that are equal to each other on $[t_o, t_1]$ except a set of points of total measure zero, are equivalent. Another example is a sampled data system (e.g., a digital controller with an A/D converter at the input) with sampling period $T$. Two different input signals that coincide only at $t = 0, T, 2T, 3T \ldots$ are equivalent to each other because the system cannot observe the values in time instants other than the sampling moments.

### 2.2.1   General Classification

Let a dynamic system has the state variable $X$. Suppose that we can define input and output signals for this system as $U$ and $Y$, respectively. The system is said to be *causal* if and only if for every $t_1 > t_0$

$U_1(t)$ equivalent to $U_2(t)$ over an interval $[t_o, t_1]$, implies that $Y_1(t_1) = Y_2(t_1)$,

where $Y_1(t_1)$ and $Y_2(t_1)$ are two outputs at the time instant $t = t_1$, obtained with inputs $U_1$ and $U_2$, respectively, $t_o$ is a fixed initial time instant, and $X(t_o)$ is fixed. The output signal is supposed to be an algebraic function of the system state. We suppose that the system state in $t_o$ is fixed.

If $U_1(t)$ is equivalent to $U_2(t)$, then we say that these functions belong to the same *class of equivalence*. Denote by **S** the set of all classes of equivalence for the dynamic system under consideration.

In the theory of dynamical systems, we can find the following classification.

**Finite automata** is a system for which the number of classes of equivalence of inputs **N** is finite. For example, an electric switch is a finite automata. A digital computer also belongs to this class, though it may have trillions of possible states at each time instant.

**Infinite automata** is a system for which **N** is infinite, but the set of possible states is enumerable. In other words, for this class of systems, the power of the set **S** is equal to the power of the set of integers. For example, a discrete-event model of one waiting line (without limitation) and one server, belongs to this class.

**Concentrated parameter systems** are those for which **N** is infinite, and the elements of the set **S** are not enumerable. For this class of systems, the power of the set **S** (its cardinal number) is equal to the power of the set of all real numbers. This is a wide class of systems that includes, among others, mechanical systems or electric circuits. However, the parameters of the systems of this class must be concentrated in specific discrete elements, like ideal springs, dampers, capacitances, etc. The most common modeling and simulation tool for concentrated parameter systems are the ordinary differential equations (ODEs).

**Distributed parameter systems**. When the power of the set **S** is greater than the power of the set of reals, the system is classified as a distributed parameter system. In other words, for such systems, the number of possible states is greater than the number of all reals. A classic example is a guitar string. Its state is a continuous three-dimensional function over the string's initial length. It is known that no one-to-one mapping between real numbers and continuous functions cannot be established, because the power of the set of such functions is greater than the power of reals. The parameters of systems of such kind cannot be concentrated in discrete components like ideal capacitors or dampers. Other examples are heat transfer problems, waves on the surface of water, and fluid dynamics. The main mathematical tool for this class of systems are the partial differential equations.

## 2.3 Linearity

A function $f(x)$ is said to be linear if it satisfies the following conditions.

$$\begin{cases} f(0) = 0 \\ f(Kx) = Kf(x) \text{ (homogenity)} \\ f(x + y) = f(x) + f(y) \text{ (additivity).} \end{cases} \tag{2.3}$$

Here, $x$ and $y$ may be real variables or points in a linear vector space, $K$ is a real constant. For example, $f = 2x$ is a linear function, while $f = 2x^2$ is not. Be careful with classifying a function as linear. For example, a function that describes a straight line on the $x$, $y$ plane, $y = f(x) = ax + b$ is an *affine function*, but it is not linear for $b \neq 0$. This is a common error to treat the affine function as linear like in [26], Chap. 7. Of course, a model described by an affine function can be easily linearized, by changing the origin of the coordinate system.

Linear systems and the above notion of linearity should not be confused with the concept of *linear space*. Recall that the space (a set with a certain structure) is lineal if there exist operators of addition and multiplication. Moreover, if A and B are elements of the space, both A+B and k×A (k is a number), must belong to the space. Obviously, the Euclidean space is lineal. The space of all continuous and bounded functions over [0, 1] with conventional adding and multiplication operators is lineal, though its elements may be non-lineal functions. However, the set of all points enclosed in a unit circle or sphere with the conventional vector summation and multiplication is not a linear space. To get more abstract linear spaces, consider, for example, a space of all hamburgers. Let's define: haburgerA+HaburgerB = a hamburger which weigh the sum of weights (norms) of both hamburgers. Let the weight of hamburgerA is equal to W kg. The result of multiplication hamburgerA by a real number k is a hamburger that weights kW kg. The origin of the space is a hamburger with weight zero. So, our space of hamburgers is a linear space because we define it just as a set of hamburgers, without imposing any limitation (note that we admit the existence of an abstract hamburger with negative weight). This space of hamburgers is not Euclidean and not metric, but it is a normed space.

An ordinary differential equation is said to be linear if both sides of the equation are linear functions with respect to the dependent variable and all its derivatives used in the expressions. For example,

$$\frac{d^2x}{dt^2} = 4x + 2\frac{dx}{dt} + 3t^2 \tag{2.4}$$

is a linear equation, while

$$\frac{d^2x}{dt^2} = 4x + 2\left(\frac{dx}{dt}\right)^3 + 3t \tag{2.5}$$

is not.

## 2.4   Ordinary Differential Equations and Models of Systems with Concentrated Parameters

An ordinary differential equation, in its general form, can be expressed as follows:

$$F(x, x^{(1)}, x^{(2)}, x^{(3)}, \ldots, x^{(n)}, t) = 0, \tag{2.6}$$

where $t$ is the independent variable (here, representing the model time), $x$ is the dependent variable, and $x^{(1)} = dx/dt$, $x^{(2)} = d^2x/dt^2$, etc. The order of the equation is equal to the order of the highest order derivative of the dependent variable. The solution to (2.6) with given initial condition for $x$ is a continuously n-times differentiable function of $t$.

The above conditions seem to be obvious, but in practice, few simulationists check them. There is a strange belief that all that is continuous, can be simulated with ODEs, and that the solutions provided by the corresponding simulation software represent or approximate the real system behavior. Unfortunately, this is not the case. In my opinion, ODEs are too primitive to be applied to global models of systems like industrial dynamics, ecology or microbiology. In Chap. 3, there is a description of other possible mathematical tools, namely *differential inclusions*.

If we can resolve the Eq. (2.6) with respect to the highest derivative of the dependent variable, then we can find the equivalent set of the first-order equations. Indeed, from (2.6), we have

$$x^{(n)} = f(x, x^{(1)}, x^{(2)}, x^{(3)}, \ldots, x^{(n-1)}, t). \tag{2.7}$$

Now, let denote $x = x_1, x^{(1)} = x_2, x^{(2)} = x_3$, etc. So,

$$\begin{cases} \dfrac{dx_1}{dt} = x_2 \\[2mm] \dfrac{dx_2}{dt} = x_3 \\[2mm] \cdots\cdots\cdots \\[1mm] \cdots\cdots\cdots \\[1mm] \dfrac{dx_n}{dt} = f(x_1, x_2, \ldots, x_n, t) \end{cases} \tag{2.8}$$

Equation (2.8) can be written in the vectorial form as follows:

$$\frac{d\mathbf{x}}{dt} = f(\mathbf{x}, t), \tag{2.9}$$

where the boldface letters denote vectors.
For example, the following equation of an oscillator

$$ax + b\frac{dx}{dt} + c\frac{d^2x}{dt^2} = 0 \tag{2.10}$$

is equivalent to the following set of two equations of the first order:

$$\begin{cases} \dfrac{dx_1}{dt} = x_2 \\[2mm] \dfrac{dx_2}{dt} = -(ax_1 + bx_2)/c. \end{cases} \tag{2.11}$$

Most of the numerical methods and their implementations use this mathematical model in the above, *canonical form*. Normally, the user is only asked to give the right-hand sides of the Eq. (2.8) and to define the initial conditions and other model parameters. The software should do the rest of the simulation task automatically.

There is a huge literature on numerical methods for ordinary and partial differential equations. Let us mention the fundamental (though antique) book of Collatz [12]. Consult also [13, 14].

Ordinary differential equations (ODEs) are used to describe dynamic systems, mainly the concentrated parameter systems. This caused the common believe that

ODE modeling is the only tool to simulate such systems. In fact, this is not true. First of all, we must be sure that the following conditions are satisfied.

1. There exist a differential equation that represents a valid model of our system.

2. If so, we must know if there exists a (unique) solution to this ODE with given initial conditions.

3. If conditions 1 and 2 are satisfied, we must check if the ODE model we use can provide solutions that satisfy the aim of the simulation task.

The use of ODEs or PDEs (Partial Differential Equations) is not the only way to describe continuous dynamic systems. A continuous model may be constructed in the form of a graphical image, with defined parameters and initial conditions. The mostly used representations are *block diagrams*, *signal flow diagrams*, and *bond graphs*.

**Block diagrams** are commonly used to describe automatic control system and instrumetantion problems.

**Signal flow diagrams** define the signal flow in the model, using arrows with well-defined direction and transfer function. Static linear and non-linear arrows are also used. See Tavangarian and Waldschmidt [30].

**Bond graphs** are used to simulate the dynamics of physical systems. In this method, we use links in the form of directed *bonds* like harpoons. The bond is associated with a corresponding *flow* (for example, velocity or electric current) and *effort* (for example, the force or electric voltage). Frequently, a bond represents the flow of energy in the real system, see Cellier [7]. The bonds are connected to *nodes*. Each node represents a balance of flows or efforts of the connected bonds.

Signal flow graphs and bond graphs are described in more detail in Sect. 2.12. The advantage of the above methods is that there are software tools that automatically generate the model ODEs, and then, the corresponding computer code, ready to execute.

## 2.5 Transfer Function

In the automatic control theory, modeling of linear dynamic systems, electronics, instrumentation, and similar fields, the dynamics of model components are frequently described by *transfer functions*, in terms of the Laplace transform (L-transform).

Let $f(t)$ be an integrable function. Recall that the *Laplace transform* (L-transform) of $f(t)$ is defined as follows:

$$\mathcal{L}\{f\}(s) = \int_0^\infty f(t)e^{-st}dt, \tag{2.12}$$

where $s$ is a complex variable, and $f(t)$ is defined on $[0, \infty]$. L-transform is a linear operator. The L-transform is a function of $s$. We will denote $\mathcal{L}(f)(*)$ as $\mathcal{L}(f)$. The theory of the Laplace transform may appear to be somewhat complicated for students and engineers. However, the transform is used to simplify and not to complicate system modeling.

Expression (2.13) defines the inverse L-transform.

$$f(t) = \mathcal{L}^{-1} f(t) = \frac{1}{2\pi i} \lim_{T \to \infty} \int_{c-iT}^{c+iT} e^{st} f(s) ds. \tag{2.13}$$

The real number $c$ is chosen so that the contour path of integration is in the region of convergence of $f(s)$.

Fortunately, the users of L-transform hardly ever have to calculate the inverse transform (2.13). Any handbook or manual on automatic control and applications of L-transform provides tables with hundreds of commonly used functions and their L-transforms already calculated. Moreover, the purpose of the models that use this tool is often the stability analysis and other properties, rather than the calculation of particular output functions.

Some of the most used properties of the L-transform are as follows.

The transform of the Dirac's pulse is $\mathcal{L}\{\delta(t)\}(s) \equiv 1$.

The transform of the unit step function $f(t) = 1(t) \equiv 1$ for $t \geq 0$ is $\mathcal{L}\{1(t)\}(s) = 1/s$.

The transform of $\frac{df}{dt}$ is $\mathcal{L}\{\frac{df}{dt}\}(s) = s\mathcal{L}\{f\}(s)$.

The L-transform of the integral $\int_0^t f(\tau) d\tau$ is equal to $\mathcal{L}\{f\}(s)/s$

The L-transform of the time-delayed function $f(t - r)$ is $e^{-rs}\mathcal{L}\{f\}(s)$

$$\mathcal{L}\{sin(\omega t)\}(s) = \frac{\omega}{s^2 + \omega^2}, \quad \mathcal{L}\{cos(\omega t)\}(s) = \frac{s}{s^2 + \omega^2}$$

To simplify the notations, we will denote the L-transform of $f(t)$ as $f(s)$.

In this section, we limit our consideration to the *linear dynamic systems* (see Sect. 2.3) that are described by linear ordinary differential equations.

Let a linear dynamic system receives an input signal $u(t)$ and produces the corresponding response as $x(t)$. The L-transforms of these signals are $u(s)$ and $x(s)$, respectively. Then, the model *transfer function* is defined as follows:

$$G(s) = \frac{x(s)}{u(s)}. \tag{2.14}$$

According to (2.14), the output of a model with transfer function $G$ is equal to $x(s) = G(s)u(s)$.

In the definition (2.14) of the transfer function, appear transforms of the signals $x(t)$ and $u(t)$ Note, however, that the transfer function does *not* depend on the input and output. It is a *property of the system model*, and not of the signals involved.

Consider the following differential equation with initial condition zero for $x$ and its derivatives.

$$a\frac{d^3x}{dt^3} + b\frac{d^2x}{dt^2} + c\frac{dx}{dt} + d\,x(t) = u(t). \tag{2.15}$$

It may be the equation of a device with input signal $u$ and output $x$.

Applying L-transform to both sides of (2.15), we have

$$as^3 x(s) + bs^2 x(s) + csx(s) + d\,x(s) = u(s). \tag{2.16}$$

From (2.16), we obtain the *transfer function* of the device:

$$G(s) = \frac{x(s)}{u(s)} = \frac{1}{as^3 + bs^2 + cs + d}. \tag{2.17}$$

This holds for signals $x(t)$ and $u(t)$ that have a corresponding L-transform. Note that $G(s)$ does not depend on $x$ or $u$. This is a property of the model, and not of the applied signals.

From the transfer function, we can go back to the corresponding differential equation.

### 2.5.1  Stability of Linear Models

Now, consider the transfer function in the general form $G(s) = \dfrac{M(s)}{N(s)}$, where $N$ and $M$ are polynomials. The *characteristic equation* that corresponds to $G(s)$ is

$$N(s) = 0. \tag{2.18}$$

Let the roots of this equation are $s_1, s_2, s_3, \ldots, s_n$, where $n$ is the order of the polynomial $N$. These roots may have real or complex values. The model described by the transfer function $G(s)$ is stable if the real parts of all roots of the corresponding characteristic equation are negative. This way, the problem of model stability is reduced to an algebraic problem. There are criteria that permit check this condition, even without calculating the roots (the Hurwitz criterion).

The consequence of the linearity of the L-transform is that the total transfer function of components connected in series is the product of the corresponding transfer functions.

### 2.5.2  Routh–Hurwitz Stability Criterion

At the end of the nineteenth century, Edward John Routh and Adolf Hurwitz proposed the criterion that determines if all roots equation $N(s) = 0$ have negative real parts, where $N(s)$ is a polynomial of order $n$.

Consider a polynomial $N(s) = a_n s^n + a_{n-1} s^{n-1} + a_{n-2} s^{n-2} + \cdots + a_1 s + a_0$.

According to the criterion, we construct the following array:

| $n$ | $a_n$ | $a_{n-2}$ | $a_{n--4}$ | $a_{n-6}$ | ..... |
|-----|-------|-----------|-----------|-----------|-------|
| $n-1$ | $a_{n-1}$ | $a_{n-3}$ | $a_{n-5}$ | $a_{n-7}6$ | ..... |
| $n-2$ | $b_1$ | $b_2$ | $b_3$ | $b_4$ | ..... |
| $n-3$ | $c_1$ | $c_2$ | $c_3$ | $c_4$ | ..... |
| $n-4$ | $d_1$ | $d_2$ | ..... | ..... | ..... |
| ........ | | | | | |

For a polynomial of order n, we must calculate n+1 rows of the array. The elements $b$ and $c$ are calculated as follows:

$$b_i = \frac{a_{n-1}a_{n-2i} - a_n a_{n-(2i+1)}}{a_{n-1}}, \tag{2.19}$$

$$c_i = \frac{b_1 a_{n-2i+1} - a_{n-1}b_{i+1)}}{b_{i+1}}. \tag{2.20}$$

Then, all the procedure repeats using the rows n−k+1 and n−k+2 while calculating row n−k. So, we have

$$d_1 = \frac{c_1 b_2 - b_1 c_2}{c_1}, \quad d_2 = \frac{c_1 b_3 - b_1 c_3}{c_1} \quad etc. \tag{2.21}$$

If there is a change of sign in the first column of the array $a_n, a_n - 1, b_1, \ldots$, then there exist roots of the polynomial $N(s)$ with a positive real part. Thus, a system described by the transfer function $G(s) = \frac{M(s)}{N(s)}$ is unstable.

It may occur that before completing the array an element with value zero appears in the first column. If so, we replace this element with $\varepsilon > 0$ and follow with calculations. After completing the array, we suppose $\varepsilon \to 0$ and examine the eventual changes of signs in the first column.

**Example**

Let $N(s) = 4s^4 + 2s^3 + s^2 + 5s + 1$.

The Routh-Hurwitz array is as follows:

| 4 | 4 | 1 | 1 |
|---|---|---|---|
| 3 | 2 | 5 | − |
| 2 | −9 | 1 | − |
| 1 | $(5*(-9) - 2)/(-9) = 47/9$ | − | − |
| 0 | 47/9 | − | − |

There are two changes of sign in the first column. So, $G(s) = \frac{M(s)}{N(s)}$ is unstable. Consult also the question and answer 2.10.

## 2.5.3  *Frequency Response*

Now, let see how a linear dynamic system with transfer function $G(s)$ responds to input signal $u(t) = \sin(\omega t)$.

In this case, the output is done as below:

$$x(s) = G(s)\frac{\omega}{s^2 + \omega^2}, \text{ that is } x(s)(s^2 + \omega^2) = G(s)\omega. \qquad (2.22)$$

Our model is lineal. This means that the response $x(t)$ to a sinusoidal input signal must also be sinusoidal, perhaps with other amplitude and a phase shift. If so, we are looking for $x(t) = g sin(t) + h cos(t)$. In terms of L-transform, it is

$$x(s) = g\frac{\omega}{s^2 + \omega^2} + h\frac{s}{s^2 + \omega^2}. \qquad (2.23)$$

From (2.22) and (2.23), we have

$$g\omega + hs = G(s)\omega. \qquad (2.24)$$

The above equations hold for any $s$, so we can substitute $s = j\omega$, where $j$ is the imaginary unity. We get

$$g\omega + hj\omega = G(j\omega)\omega, \text{ which gives } g + jh = G(j\omega). \qquad (2.25)$$

In other words, Eq. 2.24 tells us that the expression $G(j\omega)$ fully determines the output $x(t)$. $G(j\omega)$ is also called *frequency response* function. For example, observe that the gain of our device for the sinusoidal input with angular frequency $\omega$ is given by the absolute value of $G(j\omega)$, equal to $\| G(j\omega) \|$.

**Example**

Consider the model described by the following equation:

$$a\frac{d^2x}{dt^2} + b\frac{dx}{dt} + x(t) = u(t). \qquad (2.26)$$

The transfer function of this model is

$$G(s) = \frac{1}{as^2 + bs + 1}. \qquad (2.27)$$

Figure 2.1 shows the response of model (2.27) to the unit step input. The parameters are $a = 1$, $b = 0.1$. In Fig. 2.2, we can see the gain in the function of the angular frequency. The model represents a second-order low-pass filter. It can also be used as a bandpass filter with band frequency $\omega = 1.311$, $f = 0.209\,\text{Hz}$. The resulting Bode plot for (2.26) is shown on Figs. 2.2 and 2.3.

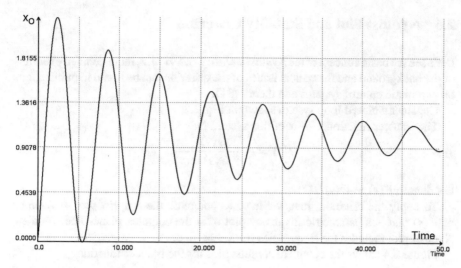

**Fig. 2.1** Response to step function

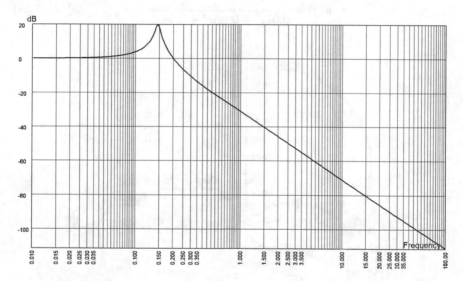

**Fig. 2.2** Spectral plot of the model (2.26)

**Fig. 2.3** A closed loop
circuit

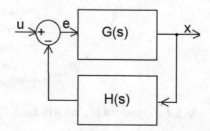

## 2.6   Nyquist Plot and Stability Criterion

Let's see the basic concept of the *Nyquist stability criterion*. A more detailed overview
of the background and theoretical issues of the criterion can be found in publications
on automatic control systems and theory [31].

Consider a closed loop system shown in Fig. 2.3

The transfer function $u \to x$ of this system is

$$D(s) = \frac{G(s)}{1 + G(s)H(s)}.$$

Let denote $F(s) = G(s)H(s)$.

To apply the criterion, first, we have to calculate the *Nyquist plot or contour*
of $F(s)$ that is a parametric frequency plot over the complex plane. The complete
Nyquist plot includes the frequencies from $-\infty$ to $\infty$.

Figure 2.4 shows the complete Nyquist plot for the transfer function

$$F(s) = \frac{1}{0.01s^3 + 0.06s^2 + 0.4s + 1}. \tag{2.28}$$

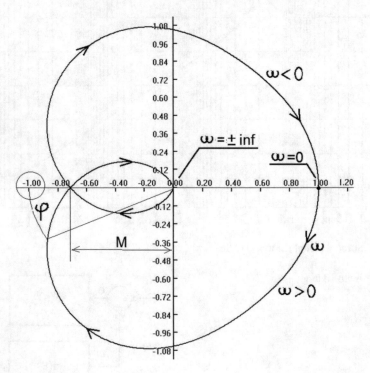

**Fig. 2.4**   Nyquist plot of a stable feedback model

Let's consider systems that are stable in open loop, that is there are no poles of $F(s)$ in the right-hand side of the complex plane.

If the Nyquist contour is closed as in Fig. 2.4, then the *Nyquist stability criterion* states that the necessary and sufficient condition for the feedback system $D(s)$ to be stable is that the complete Nyquist diagram of function $F(s)$ does not touch nor encircle the critical point $(-1, j0)$.

The value $1/M$ (Fig. 2.4) is the *gain margin*. This means that the system becomes unstable if we multiply the feedback gain by $1/M$. Similarly, the angle $\varphi$ is the phase margin. If we decrease the phase angle of $F(s)$ by $\varphi$, then the system loses stability (phase angle decreases clockwise).

Consult Willems [31] for more detailes of the Nyquist criterion..

## 2.7 Analog Computer Models

A typical example of analog (physical) simulation is a wind-tunnel test on a scaled-down physical model of an airplane. Another way to do analog simulation is to look for an analogy between one physical model and another, for example, a mechanical system and an electric circuit. A simple electronic circuit with one operational amplifier, one resistance, and one capacitor can realize the operation of mathematical integration. As a consequence, it is possible to solve differential equations using combinations of such circuits. The advanced devices with many such circuits and variable interconnections between them are called analog computers.

In the early 40 s and 50 s, analog computers were commonly used to simulate continuous dynamic systems, mainly automatic control systems, mechanical systems, and similar. During the last decades, analog computers have been losing importance. However, we should remember that analog computers are truly continuous parallel differential equation solvers. As stated in the Introduction, continuous simulation of digital computers is a much more "artificial" approximation of real continuous problems.

Figure 2.5 shows an analog circuit that satisfies the following equation:

**Fig. 2.5** Analog model of an oscillator

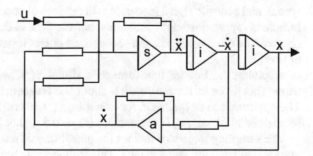

**Fig. 2.6** Analog integrator

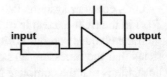

$$d^2x/dt_2 = -u - x - adx/dt,$$

where $u(t)$ is an external input and $x(t)$ is the model response. Operational amplifiers marked with i are integrators, the amplifier marked with $s$ is a summator and the amplifier marked with $a$ is an inverting amplifier. All amplifiers are supposed to have negative gain. The rectangles represent resistances. The analog integrator is realized by the circuit shown in Fig. 2.6.

Analog computers are no longer used to solve differential equations. However, the analog models that simulate given dynamic elements of transfer functions are still used in automatic control, instrumentations, and analog signal filtering. The old, vacuum bulbs analog computer with about 20 operational amplifiers was a big device weighing tens of kg. A contemporary IC operational amplifier weight is no more than several milligrams.

## 2.8  Z-transform

This kind of model is mentioned in the preset chapter, because the models considered below are still continuous with respect to the model state. The discretization refers only to the time variable.

In automatic control and instrumentation, the data taken from control objects or from the environment, need not be gathered continuously. For example, thermal objects like heaters may have time constants up to hours or days. Such objects can be observed with sufficient accuracy by taking the measurements one per 10 seconds or one per 10 minutes. In automatic control, such circuits are called *sampled data systems*. Recently, when the data processing can be done with extremely high speed, we can take the data, for example, with a frequency of more than 40 kHz, like in speech and acoustic signal processing. In devices of such kind, the value of the data sample is (approximately) continuous, but the time is discrete. This means that we can treat the observed signal as a sequence of pulses instead of a continuous function of time.

Applying the Fourier transform to a signal $x(t)$, we can get the signal *spectrum*. This gives us the information about the frequencies contained in the signal. The spectrum can be limited if, for example, $x(t) = sin(5t) + sin(8t)$, or unlimited (contained in $[0, \infty]$) if, for example, $x(t) = 0.5 + 0.5sign(sin(\omega t))$.

The sampling frequency defines the possibility of retrieving the original, continuous signal from the sampled data. This frequency is limited from below, according to the Nyquist–Shannon sampling theorem [22]: *"If a function $x(t)$ contains no frequencies higher than B hertz, it is completely determined by giving its ordinates at a series of points spaced $1/(2B)$ seconds apart."*

This is why the sampling frequency for digital processing of acoustic signals is normally set greater than 40 kHz, and in the processing of heart acoustic signals, the sampling frequency should not be below 250Hz. If the frequencies contained in the original signal are greater than the *Nyqyuist frequency B*, then some undesired phenomena may occur, like *aliasing*. This may produce false signals. To avoid aliasing in digital signal filters, additional simple analog low-pass filters may be added to the device input.

The sampled data systems should be analyzed using the Z-transform. Considering the signal $x_n$ as a sequence of pulses, we can define *Z-transform* as follows:

$$\mathcal{Z}\{(x_n)\} = \sum_{n=-\infty}^{\infty} x_n z^{-n}, \tag{2.29}$$

where $x[n]$ is a sequence $x_1, x_2, \ldots, x_n \ldots$ and $z$ is a complex variable. It is supposed that the samples are taken at time instant $T, 2T, 3T, \ldots$, where $T$ is the step of time-discretization. The above transform is called *two-sided*. In the *one-sided transform*, the sum in (2.29) runs for non-negative values of $n$.

In (2.29), $z$ is a complex variable, $z = Ae^{j\phi}$, where $j$ is the imaginary unit, $\phi$ is a parameter called *angle*, and A is a real number. Like the L-transform, Z-transform has also its inverse. We will not discuss this inverse here, because the users, such as control systems or signal processing engineers, hardly ever need to calculate it. We also have $z = e^{sT}$, where $s$ is the Laplace transform variable.

Z-transform is a linear operation. We will not give here an overview of the theory of the transform, discussing only very elemental properties that may be easily used in sample data system analysis. Perhaps the most important property is as follows: Given a sequence $\{y_n\}$ such that $y_n = x_{n-k}$, we have

$$\mathcal{Z}\{(y_n)\} = z^{-k}\mathcal{Z}\{(x_n)\}. \tag{2.30}$$

In other words, the operation of delaying the sample by $kT$ corresponds to multiplying the Z-transform by $z^{-k}$ (T is the sample period)

We should remember that the sampled data system is a thing quite different from a continuous-time system. We do not pass with the sample step to zero, so $T$ is always finite and fixed. However, the behavior of sample data model may be similar to a continuous system. There are several methods that permit to find a continuous model that corresponds to a given Z-transform model and vice versa. However, such mappings are hardly ever unique.

Let's see one, perhaps the simplest method to get a Z-transform version of a model given in the form of a (L-transform) transfer function $G(s)$. Let our model has the transfer function as follows:

$$G(s) = \frac{1}{a_n s^n + a_{n-1} s^{n-1} + \ldots + a_0}. \tag{2.31}$$

If the device input is $u(s)$ and output $x(s)$, $\frac{x(s)}{u(s)} = G(s)$. So, we get the following corresponding differential equation model:

$$a_n \frac{d^n x}{dt^n} + a_{n-1} \frac{d^{(n-1)} x}{dt^{(n-1)}} + a_{(n-2)} \frac{d^{(n-2)} x}{dt^{n-2}} + \ldots + a_0 x(t) = u(t). \tag{2.32}$$

Now, replace the differential Eq. (2.32) with a difference equation with time-step $T$, where $x_k = x(kT)$. The derivative of $x$ is approximated with $\frac{x_k - x_{k-1}}{T}$. The higher derivatives can be converted using the following formula (backward difference):

$$\frac{\nabla_h^n x(t)}{T^n} = \frac{\sum_{i=0}^n (-1)^i \binom{n}{i} x(t - iT)}{T^n}$$

For example, the first-order inertial object $G(s) = 1/(1 + \theta s)$ will convert into $x(t) + \theta dx/dt = u(t)$ that gives the time-discrete version: $x_k + \theta \frac{x_k - x_{k-1}}{T} = u_k$ ($\theta$ is the time constant of the object). Now, recall that the one-step delay corresponds to $z^{-1}$. So, we have

$$x(z) + \frac{\theta}{T} x(z)(1 - z^{-1}) = u(z), \tag{2.33}$$

where $x(z)$ and $u(z)$ are Z-transforms of $x$ and $u$, respectively.

From (2.33), we can calculate

$$G(z) = \frac{x(z)}{u(z)} = \frac{1}{1 + \frac{\theta}{T}(1 - z^{-1})}. \tag{2.34}$$

This way, we obtain the transfer function in terms of Z-transform that corresponds to $G(s)$.

For example, the transfer function of a continuous integrator $1/s$ with $T=1$ will be

$$G(z) = \frac{1}{(1 - z^{-1})}.$$

Using this form of the integrator, a simple Z-transform version of a PI controller will be as follows:

$$G(z) = K \left( 1 + \frac{1}{T_i(1 - z^{-1})} \right). \tag{2.35}$$

Figure 2.7 shows the response of a sampled data PI controller (2.35) to a rectangular input. Sampling period equal to one, $K = 1$, $T_i = 5$. The input is equal to one for $5 < t \leq 30$, zero otherwise. It can be seen that this is a quite good approximation of a corresponding continuous controller.

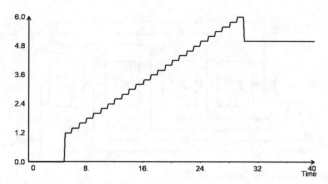

**Fig. 2.7** Response of a sampled-data PI controller

As mentioned earlier, the conversion from $G(s)$ to $G(z)$ is not unique. Using the forward difference scheme, we will obtain a different expression. There are also other methods of conversion from Laplace to Z-transform, for example, the *pole-zero mapping* (consult Ifeachor and Jervis [19]), and Sect. 2.8.1.

For higher order objects, the transform function can have the form:

$$G(z) = \frac{b_m z^{-m} + b_{m-1} z^{-m+1} + \cdots + b_0}{a_n z^{-n} + a_{n-1} z^{-n+1} + \cdots + a_0}. \qquad (2.36)$$

If we multiply the numerator and denominator of (2.36) by $z^n$, the denominator converts into a polynomial of order $n$ with respect to $z$. The object described by $G(z)$ is stable if all the roots of this polynomial are contained in the unit circle on the (complex) z-plane. In the expressions of numerator and denominator, each real pole or zero at $a$ converts into $(z - a)$. Recall that each pair of complex poles/zeros at $Ae^{\pm j\phi}$ converts in the term $z^2 - 2Az\cos(\phi) + A^2$.

From the above simple introduction, we can see how useful is the Z-transform in sampled-data control systems and digital signal processing. Complex devices can be constructed using simple delay components $z^{-1}$ (one-step delay). This permits the same hardware to be used for time-delay acoustic signal devices, active filers, and other complicated signal processing systems, changing only the implemented software.

The implementation of the one-step delay element $z^{-1}$ on digital or hybrid devices is simple, compared to the analogic integrator device. A small IC may contain thousands of such elements. This permits to construct low-cost circuits that can realize complicated Z-transform functions. For example, consider the following transfer function:

$$G(z) = \frac{x(z)}{u(z)} = \frac{b_3 z^{-3} + b_2 z^{-2} + b_1 z^{-1} + b_0}{a_4 z^{-4} + a_3 z^{-3} + a_2 z^{-2} + a_1 z^{-1} + a_0}. \qquad (2.37)$$

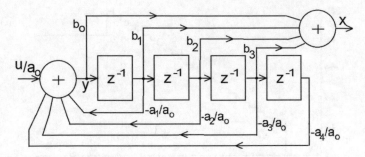

**Fig. 2.8** Circuit with transfer function (2.37)

The circuit that provides this transfer function may be constructed with four one-step delay elements as in Fig. 2.8. It can also be coded to be executed on a computer.

The signal links have gains, as indicated in the scheme. Signal $y$, according to the first summator, is as follows:

$$y = \frac{1}{a_0}u - y\frac{a_4}{a_0}z^{-4} - y\frac{a_3}{a_0}z^{-3} - y\frac{a_1}{a_0}z^{-1} - y\frac{a_1}{a_0}z^{-1}.$$

Thus, we get

$$y = \frac{u}{a_4 z^{-4} + a_3 z^{-3} + a_2 z^{-2} + a_1 z^{-1} + a_0}.$$

From the scheme, we have also

$$x = b_0 y + b_1 y z^{-1} + b_2 y z^{-2} + b_3 y z^{-3}.$$

From the last two equations, we can calculate $x(x)/u(z)$. This gives us the desired transfer function (2.37).

### 2.8.1  Matched Pole-zero

The *matched pole-zero* is another method for obtaining transfer function $G(z)$ from the continuous version of the modeled object [19]. Let $G(s)$ has the form $G(s) = M(s)/N(s)$, where $M$ and $N$ are polynomials. Suppose that the order of $N$ is greater than the order of $M$ and that we know the values of the roots of $M$ (zeros of $G$) and of $N$ (poles). The procedure is as follows:

1. Each finite pole or zero of $G$, $s = -a$, is replaced with a pole in $z = e^{-aT}$.
2. Each infinite zero of $G$, is replaced with zero at $z = -1$.
3. Overall gain of the object $G(z)$ must be adjusted, to match the gain of the continuous version, for a selected frequency. Normally, we require that the gains match for the steady state, that is, for the frequency equal to zero.

Here, $T$ is the sampling time interval.

Consider, for example, the following object:

$$G(s) = \frac{s+4}{(s+2)(s+5)}. \tag{2.38}$$

Poles of this object are $s_{1,N} = -2$, $s_{2,N} = -5$. Finite zero of $G(s)$ is $s_{1,M} = -4$. $G(s)$ has also a zero in infinity. According to the matched pole-zero method, the corresponding sample-data transfer function is as follows:

$$G(z) = K \frac{(z+1)(z - e^{-4T})}{(z - e^{-2T})(z - e^{-5T})}, \tag{2.39}$$

where $K$ is a constant.

Now, we must adjust $K$. Let assume that the gain of $G(s)$ and $G(z)$ is the same for input frequency $\omega$ equal to zero. So, we substitute $s = j\omega = 0$ and, consequently, $z = e^{j\omega T} = 1$. From (2.38) and (2.39), we have

$$\frac{4}{10} = K \frac{2(1 - e^{-4T})}{(1 - e^{-2T})(1 - e^{-5T})}. \tag{2.40}$$

Finally,

$$K = 0.2 \frac{(1 - e^{-2T})(1 - e^{-5T})}{(1 - e^{-4T})}. \tag{2.41}$$

Observe that $G(s)$ and $G(z)$ cannot be strictly equivalent. $G(s)$ is a function of $s$, and $G(z)$ is a function of $z$ and of the sampling time $T$. Moreover, there may be various forms of $G(z)$ derived from the same $G(s)$, depending on the method used.

Note also that, by sampling a signal $x(t)$, we lose the information about the signal that is provided by the values of $x$ at a time not equal to $nT$ (n integer). For example, the signal $x(t) = cos(2\pi t/T)$ sampled at $t = 0, t = T, t = 2T, \ldots$, gives the same output from the sampled data object, as the signal $x(t) \equiv 1$.

The above considerations represent only a short introduction to modeling sampled data systems with Z-transform. For further reading consult, for example [19].

## 2.9  Non-linear Models and Stability

Non-linear models do not satisfy the linearity criterion defined in Sect. 2.3. Let us focus on models described by non-linear ordinary differential equations. First of all, note that **non-linear model does not have a transfer input–output function**. While integrating non-linear equations, we can find some computational difficulties. However, perhaps the most difficult problems are those of the stability of non-linear systems. Recall that a linear system can be stable or not. This is the property of the set of model equations or of the transfer functions. A non-linear system is characterized by several kinds of "stabilities," depending on which stability definition we use.

The non-linear stability is widely treated in works on control theory and automatic system engineering [4, 11, 17, 18, 23], and it is not the main topic of this book. Here, we recall only the general concepts.

To focus on some general forms of model, we will consider a dynamic system described by the following equation:

$$\dot{\mathbf{x}} = \mathbf{f}(\mathbf{x}, \mathbf{u}, t), \tag{2.42}$$

where $t \geq 0$, $\mathbf{x} \in R^n$, $\mathbf{u} \in R^m$, and $\mathbf{u}$ is a bounded, integrable external control function.

### 2.9.1  BIBO Stability

This is *Bounded Input, Bounded Output* stability. In few words, we require that if we apply a bounded external excitation to the system, the corresponding response be bounded also. The term "bounded" means that we measure the signal using a certain *norm* of the function. In the simplest case, it may be the maximum of the absolute value of the (scalar or vectorial) signal over the time. This type of stability may not be very practical in certain applications. For example, if the norm of the input signal is small, and the norm of the corresponding response is very big, then the system is still considered BIBO-stable, though it may overcome permissible limits.

### 2.9.2  Lyapunov Stability

Let $\mathbf{u}$ be a fixed control, $t = t_0$, $\mathbf{x}(t_0) = \mathbf{x}_1$, and $\mathbf{x}_1$ be an equilibrium point. If for any $\varepsilon > 0$ there exists a $\delta(\varepsilon, t_0) > 0$, such that

$$\| \mathbf{x}_1 - \mathbf{x}(t_0) \| < \delta \Rightarrow \| \mathbf{x}_1 - \mathbf{x}(t) \| < \varepsilon \ \forall \ t > t_9, \tag{2.43}$$

then we say that the system is *stable in the sense of Lyapunov* at $\mathbf{x}_1$.

### 2.9.3  Asymptotic Stability

Let the conditions of Sect. 2.9.2 hold. If a constant $\delta(t_0) > 0$ exists, such that

$$\| \mathbf{x}_1 - \mathbf{x}(t_0) \| < \delta \Rightarrow \| \mathbf{x}_1 - \mathbf{x}(t) \| \to 0 \ \text{ when } t \to 0 \tag{2.44}$$

then the system is it asymptotically stable at the point $\mathbf{x}_1$.

### 2.9.4 Orbital Stability

Many dynamic systems never reach any final steady state and enter into oscillations. Such oscillations, shown on the phase plot, for example, on the $x_k, x_j$ plane, look like closed contour on which the state coordinates move. If we apply an external disturbance to the system, the trajectory may go out from the cycle, or return to it after some time interval. In the last case, we say that the system is *orbitally stable* with respect to the cycle. We will not enter here in the strict mathematical definition but illustrate a possible stable cycle (orbit) on the phase plane.

Consider a fourth-order system, described by the following equation:

$$
\begin{cases}
dx_1/dt = ax_1 + bx_1x_2 + cx_1x_4 \\
dx_2/dt = dx_2 + ex_1x_2 + fx_2x_4 \\
dx_3/dt = gx_3 + hx_2x_3 + ix_2x_4 + jx_3x_4 \\
dx_4/dt = kx_4 + lx_1x_4 + mx_2x_4 + nx_3x_4
\end{cases}
\tag{2.45}
$$

This is a Lotka–Volterra equation for a prey–predator system with four interacting species. Figure 2.9 shows the system trajectory projected into the plane $(x_1, x_2)$. It can be seen that after the initial "warm-up" period, the system reaches a (perhaps) stable oscillation cycle. Model parameters are as follows:

$a = 0.1, b = -0.002, c = -0.0001, d = -0.3, e = 0.001, f = -0.000015,$
$g = 0.003, h = 0.00001, i = 0.000012, j = -0.000012, k = -0.01, l = 0.00002,$
$m = 0.00004, and n = 0.000015.$ The initial conditions are $x1 = 500, x2 = 50, x3 = 100, and x4 = 20,$ simulation final time equal to 2000.

The *Jacobian* of system (2.42) is defined by

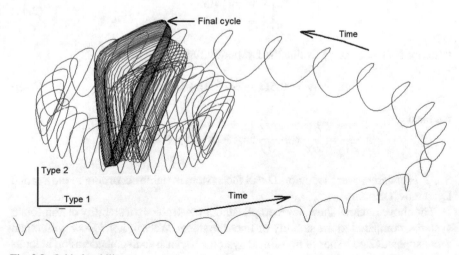

**Fig. 2.9** Orbital stability

$$J(\mathbf{x}) = \frac{\partial \mathbf{f}}{\partial \mathbf{x}}. \tag{2.46}$$

There are two stability criteria known as *Lyapunov theorems*. Let function $\mathbf{f}$ be continuously differentiable. The *first Lyapunov Method* is given by the following theorem.

**Lyapunov Theorem 1**. Let $\mathbf{x}^*$ be an equilibrium point of (2.42). If all the eigenvalues of $J$ have a negative real part, then the system (2.42) is asymptotically stable around $\mathbf{x}^*$.

We will say that a function $g$ is of class $\mathcal{K}$, if $g(t_0) = 0$, $g(t) > 0$, and $g(t)$ is continuous and nondecreasing for all $t > t_0$.

Now, consider the following scalar function, named *Lyapunov function*:

$$\begin{cases} (a) \ V(0, t_0) = 0 \\ (b) \ V(\mathbf{x}, t) > 0 \ \forall \ \mathbf{x} \neq 0, \ t \geq t_0 \\ (c) \ \alpha(\| \mathbf{x} \|) \leq V(\mathbf{x}, t) \leq \beta(\| \mathbf{x} \|) \ \forall \ t \geq t_0, \\ (d) \ \dot{V}(\mathbf{x}, t) \leq -\gamma(\| \mathbf{x} \|), 0 \ \forall \ t \geq t_0, \end{cases} \tag{2.47}$$

where $\alpha$, $\beta$ and $\gamma$ are functions of class $\mathcal{K}$. Suppose that system (2.42) has an equilibrium at the point $x_0 = (0, 0, \ldots, 0)$.

**Lyapunov Theorem 2**, of the **the Second Method of Lyapunov** states that the system (2.42) is globally, uniformly (with respect to $t_0$), and asymptotically stable if the function $V(\mathbf{x}, t)$ exists.

The condition (d) of (2.43) is satified if $d(V)/dt < 0$.

Example. Consider the following system (undamped pendulum):

$$\begin{cases} \dot{x} = -sin(y) \\ \dot{y} = x \end{cases} \tag{2.48}$$

where $y \in D = (-\pi, \pi)$ Define the Lyapunov function as follows:

$$V = 0.5(1 - cos(y)) + 0.5x^2, \tag{2.49}$$

we have

$$\frac{dV}{dt} = \frac{\partial V}{\partial x}\dot{x} + \frac{\partial V}{\partial y}\dot{y} = -xsin(y) + 0.5xsin(y)$$

So, $\dot{V}$ is negative over the region $D$ and the system is stable according to the Second Lyapunov Theorem.

The above remarks show how complicated is the problem of stability of non-linear systems, compared to the stability of linear systems. We will not follow with other, more sophisticated issues of non-linear systems because such consideration belongs rather to advanced mathematics and not modeling and simulation.

**Fig. 2.10** A control circuit
with non-linear element

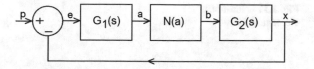

There are several practical methods of stability analysis, developed for non-linear control feedback systems. Let us comment on one of them, perhaps the most representative and easily applicable. The method of *Describing Function* has been developed in the USSR in the late 1930s, by Nikolay Mitrofanovich Krylov and Nikolay Bogoliubov [20].

Figure 2.10 shows a feedback control system with non-linear element. Signal $p$ is the set point, $e = x - p$ is the control error and $x$ is the controlled process output. $G_1(s)$ is the transfer function of the controller, and $G_2$ is the transfer function of the controlled process. $N(a)$ represents the non-linear part of the circuit. This may be a saturation, a friction non-linearity, hysteresis, or another non-linear element.

The describing function "describes" the non-linear element, with respect to the response to a sinusoidal input. Suppose that $a(t) = A sin(\omega t)$. The output of the non-linear element needs not be sinusoidal. Let's express it as the corresponding Fourier series.

$$b(t) = B_0 + A_1 cos(\omega t) + B_1 sin(\omega t) + A_2 cos(2\omega t) + B_2 sin(\omega t) + A_3 cos(3\omega t)$$
$$+ B_3 sin(3\omega t) + \cdots\cdots . \tag{2.50}$$

We can suppose $B_0 = 0$ because the non-linearities are in most cases symmetrical. The main assumption of the describing function is that the elements $G_1$ and $G_2$ (or at least one of them) are low-pass filters. It is also required that the Fourier series of $b(t)$ converges.

So, we can use the approximation of $b(t) = A_1 cos(\omega t) + B_1 sin(\omega t)$ instead of the original signal. This can also be expressed as $b(t) = C_1 sin(\omega t + \phi_1)$, where $\phi_1$ is the phase shift of the first harmonic of $b(t)$. Using the complex representation of phase-shifted signal, we have

$$B\angle\phi_1 = B_1 + jA_1 = \sqrt{B_1^2 + A_1^2}\angle tan^{-1}(\frac{A_1}{B_1}). \tag{2.51}$$

The *describing function* is defined as follows:

$$N = \frac{B}{A}\angle\phi_1. \tag{2.52}$$

Note that $N$ is a complex-valued function that depends on the amplitude $A$ of the input signal. The absolute value of function $N(A)$ is the gain of the non-lineal element for the first harmonic of the output signal, and its angle is the phase-shift of the first harmonic. Remember that we neglect higher harmonics of signal $b(t)$. So,

**Fig. 2.11** Non-linear
element: saturation

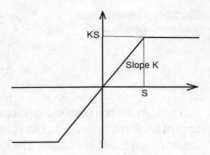

$N(A)$ provides an information similar to the frequency response function $G(j\omega)$. This property is used to analyze the stability of the system, using the Nyquist stability criterion [15]. If the Nyquist plot (in the function of the frequency) and the plot of the describing function (in the function of amplitude) are both plotted at the same complex plane, then the intersection point of these two plots (if any) gives us the information about a possible oscillation, namely the frequency and the amplitude.

It may appear that the describing function may be difficult in practical applications because of the need to calculate the first terms of the Fourier transform. However, this is not true. Most of the non-linearities that can appear in practical applications have the corresponding describing functions already calculated and available in control engineering handbooks. See, for example, the characteristic of the saturation element as in Fig. 2.11.

The describing function for this element is

$$\begin{cases} N = \dfrac{B}{A} \angle 0^o \\[2mm] = \dfrac{2K}{\pi} \left[ sin^{-1}\left(\dfrac{S}{A}\right) + \dfrac{S}{A}\sqrt{1 - \left(\dfrac{S}{A}\right)^2} \right] \angle 0^o. \end{cases} \qquad (2.53)$$

Using the Nyquist criterion, we simply suppose that the total gain (frequency response) of the feedback loop is $M(A, j\omega) = G_1(j\omega)N(A)G_2(j\omega)$. If the Nyquist plot for function $G_1 G_2$ intersects with the plot of $-1/N(A)$ (as function of $A$), then the intersection point gives us the information about $\omega$ (from $G_1 G_2$) and $A$ (from $N(A)$). For more rigorous statement, consult [15]. Note that the shape of the plot depends on the amplitude A. So, for different A we can obtain different stability properties (Fig. 2.12).

**Fig. 2.12** An example of
Nyquist plot and describing
function on the complex
plane ($G = G_1 G_2$)

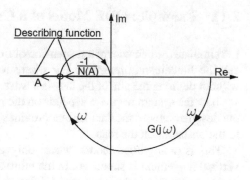

## 2.10  Stiff Equations

Stiffness occurs in a problem where there are two or more very different time scales, for example, in systems that have very fast and very slow parts interacting with each other. Consider the following set of equations:

$$\begin{cases} dx_1/dt = 998x_1 + 1998x_2 \\ dx_2/dt = -999x_1 - 1999x_2 \\ x_1(0) = 0, \ \ x_2(0) = 0 \end{cases} \tag{2.54}$$

The solution to these equations is

$$x_1(t) = 2e^{-t} - e^{-1000t}, \ \ x_2(t) = -e^{-t} + e^{-1000t}.$$

To integrate this system of equations, we must define the step size of any numerical method, as $h << 0.001$ if we want the method to be stable. However, the term $e^{-1000t}$ disappears very quickly when $t$ grows. One could suppose that for $t$ between, say 0.1 and 1 we could apply a greater integration step because the fast term does not influence the solution. Unfortunately, it is not the case, because the integration process becomes unstable even if this term is negligible. This problem is called *stiffness* and provokes serious difficulties in system simulation.

The most popular methods for stiff equations are:

* Generalizations of the Runge–Kutta method, of which the most useful are the Rosenbrock methods, and their implementations.

* Generalization of the Burlisch–Stoer method, due to Bader and Deufhard [2].

* Predictor–corrector methods.

For more detail, consult Bader [2].

## 2.11  Example: ODE Model of a Car Suspension

Let simulate the behavior of a suspension of a car. First of all, the simulationist should ask the fundamental question: For what? Remember that to simulate something without defining the aim of the task is a waste of time. In our case, the problem is to see how the system response depends on the parameter of the damper. This can help the designer choose the damper that provides an acceptable response for the vehicle to the obstacles on the road.

This is a simplified model, where only one wheel is considered and only the vertical movement is simulated. In the more advanced models of vehicle dynamics, we must simulate all the forces the vehicle receives and look for the three-dimensional model of the system and the dumper non-linearity. To start with a simple academic example, the following model can be quite illustrative.

Suppose that the car moves forward with a constant horizontal velocity $V$. We will take the car position as the reference and consider all variables as differences between their actual values and the initial equilibrium state. Figure 2.13 shows the mechanical scheme of the model. The variables are as follows.

$y$—vertical movement of the wheel;

$x$—vertical movement of the vehicle;

$F_1$—the force determined by the compliance of the tire;

$F_2$—the force of the suspension spring;

$F_a$—the force produced by the damper;

$u$—an external excitation, due to the shape of the road.

All the above variables are functions of time and all represent vertical movements and forces. $M_1$ and $M_2$ represent the mass of the wheel (together with the moving part of the suspension) and the mass of the car, respectively ($M_2$ being 1/4 of the car's total mass).

To derive the model equations, we must use the force balance for the two masses, including the dynamic forces. This provides the following equations:

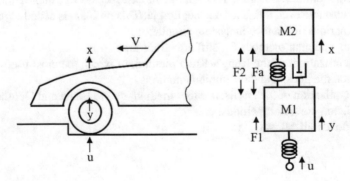

**Fig. 2.13**  A car suspension

$$\begin{cases} F_1 - F_2 - F_a - M_1 d^2 y/dt^2 = 0 \\ F_2 + F_1 - M_2 d^2 x/dt^2 = 0. \end{cases} \tag{2.55}$$

The forces of the two springs and of the damper depend on the corresponding changes in the spring and damper length, so

$$F_1 = K_1(y - u), \quad F_2 = K_2(x - y), \quad F_a = K_a(dx/dt - dy/dt).$$

Note that the damper force depends on the velocities and not on positions. The above forces are supposed to be linear with respect to the positions and velocities. The next version of the model might include non-linear functions for the forces. This would not complicate significantly our simulation task; the only necessary change would be to replace the above expressions with non-linear functions.

As the most convenient form of an ODE model is a set of equations of the first order, let us redefine the equations using the following notation:

$$x_1 = x, \quad x_2 = dx/dt, \quad x_3 = y, \quad x_4 = dy/dt.$$

After substituting the new variables into (2.55) and reordering the equations, we obtain the following:

$$\begin{cases} dx_1/dt = x_2 \\ dx_2/dt = [K_2(x_2 - x_1) + K_a(x_4 - x_2)]/M_2 \\ dx_3/dt = x_4 \\ dx_4/dt = [K_1(u(t) - x_3) - K_2(x_3 - x_1) - K_a(x_4 - x_2)]/M_1 \end{cases} \tag{2.56}$$

The last set of equations can be used to simulate our system. We obtain four equations, and the state vector of the model is $x = (x_1, x_2, x_3, x_4)$. This is correct because the movement equation for each mass is of the second order.

Now, we must decide what to do with our mathematical model. A beginner could choose the following procedure. First, find a numerical algorithm to solve the equations, for example, Runge–Kutta–Fehlberg. The algorithm can be found in any book on numerical methods. Then, prepare the corresponding code, in Basic, Pascal, Fortran, C, or other programming languages. Insert the code of our model (Eq. (2.56)) in the program, compile, and run it. This may result in a correct simulation program and provide good results, but in most cases, it is simply a waste of time. The same task can be completed 20 times faster while using an appropriate simulation language. In any directory of simulation software, like the Directory of Simulation Software of the Society for Computer Simulation you can find hundreds of simulation languages and packages, at least half of them for continuous ODE models. A good ODE simulation package should only ask you to type the right-hand sides of the model equations and to define model parameters and initial conditions. The rest should be done automatically, providing all needed reports, trajectory plots, etc.

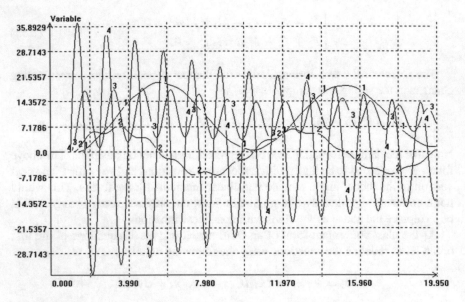

**Fig. 2.14**   Car suspension plots

In the following figures see an example of simulation results. The plots were generated by the ODE module of the BLUESSS (see Sect. 8.3) simulation system. The integration algorithm was Runge–Kutta–Fehlberg of order 5, with about 2000 time-steps. The system parameters are as follows:

$$M_1 = 50 Kg, \, M_2 = 200 Kg, \, K_1 = 1000 N/cm, \, K_2 = 100 N/cm.$$

The external excitation $u(t)$ was a step function with amplitude 10cm, starting at $t = 1$ s. Note that if you use the SI unit system, the mass must be given in Kg and the force in Newton.

The aim of our task is to see how the damper parameter $F_a$ affects the system performance. The value used for the simulation of Fig. 2.14 is too small and results in a highly oscillatory response. Figure 2.15 shows the result of a simulation experiment where $F_a$ changes automatically from 5 to 80 in 25 steps.

The trajectory is automatically simulated for each value of $F_a$. The 3D plot of Fig. 2.15 shows the changes in the trajectory shape. The vertical coordinate is the car vertical position, the axis from left to right is the model time and the other one (marked P) is the value of $F_a$. Running the program several times the designer can choose a satisfactory value for $F_a$.

Most of the dynamic systems are subject to stochastic disturbances. The same package permits other kinds of experiments, where one or more inputs are random signals. Figure 2.16 shows the trajectory of our model (car vertical position), where the excitation $u$ is as before, plus a uniformly distributed random value with mean zero and amplitude 20, changing at each integration step of 0.01 s. The plot of Fig. 2.16

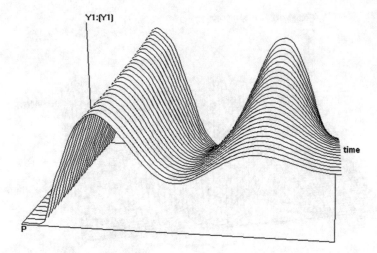

**Fig. 2.15**  A set of system trajectories with different values of the damper parameter (screenshot)

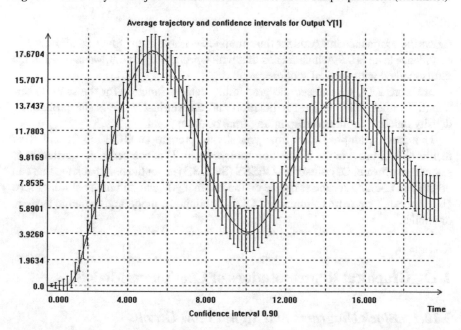

**Fig. 2.16**  System trajectory and confidence intervals in the case of a stochastic disturbance. Damper parameter $F_a = 30$.

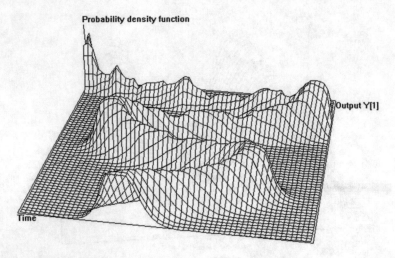

**Fig. 2.17**  The probability density function for car elevation

shows the confidence intervals for the simulated variable as the function of time. The confidence level is 0.9, which means that with a probability of 0.9, we are within the limits marked with vertical sections.

In Figure 2.17, we can see the probability density function for the same experiment. "Output Y[1]" is the car position and the vertical value is the probability density function for the corresponding time position point.

The above example shows how you should prepare an ODE model and what results should provide the simulation tool you use. This can serve as a criterion to select the software. Of course, BLUESSS (Sect. 8.3) is not the only package that can do the job. Looking for the software on the Internet you can find many similar tools. Use the search keywords simulation, ODE, continuous, modeling, dynamic system, or similar.

## 2.12  Graphical Representations of Continuous Models

### 2.12.1  Block Diagrams and Signal Flow Graphs

*Signal Flow Graphs* (SFGs) can be used to represent the dynamics of a modeled system, instead of differential equations. The very traditional way to simulate dynamic systems has been to obtain the system equations, prepare the corresponding code in a programming language and run the simulation. However, observe that a simulationist needs not be a mathematician or a programmer. The simulation software and new modeling methods make it possible to eliminate both mathematics and coding from the modeling and simulation tasks. What the simulationist must do is to

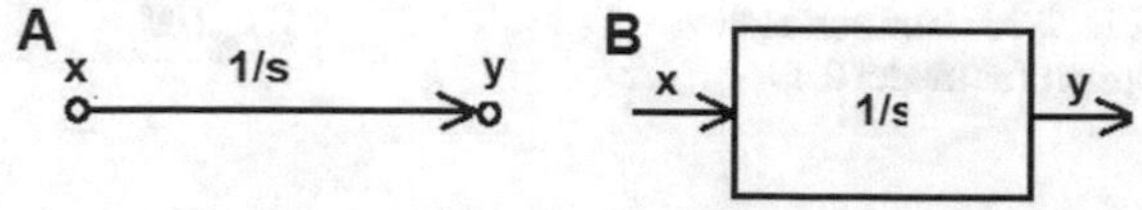

**Fig. 2.18**  Signal flow graph
and block diagram

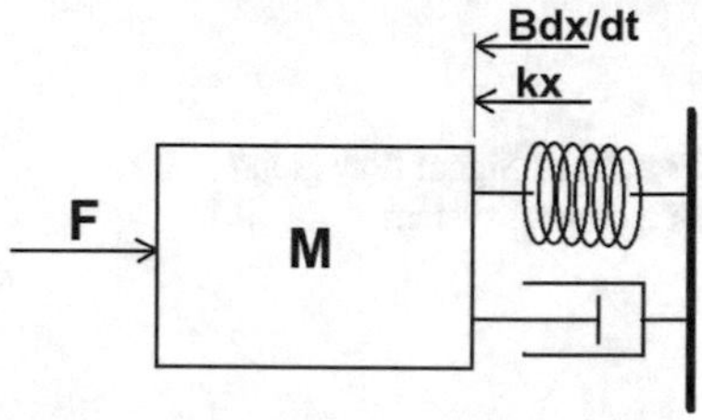

**Fig. 2.19**  A mechanical
model

understand the structure and the dynamics of the modeled system and to be able
to describe it precisely enough to be interpreted by a computer. Both signal flow
graphs, block diagrams, and bond graphs (see the next section) are graphical representations of dynamic system models. If the model is described in such a graphical
way, the simulation software should automatically generate model equations and the
corresponding code and run the simulation. A similar model representation can be
created using *block diagrams* that are commonly used in modeling automatic control
circuits.

By the SFG we mean a network composed of nodes and directed links. Nodes
represent signals and links represent transfer functions. The direction of a link shows
which signal is the input to the link (the cause) and which is the output (the result).
Figure 2.18 shows a graph that describes an integrator. Signal $y$ depends on the signal
$x$ and not vice versa. The transfer function (in terms of Laplace transform) for the
link between A and B is $G(s)=1/s$. This means that $y$ is the integral of $x$. Part B of
the figure shows the equivalent block diagram. If more than one link enters (points
to) the node, then the effect of them is the sum of the corresponding signals (at the
node). A node with no entering links is called a *source node*.

Consider a simple mechanical system composed of a spring, a mass, and a damper,
shown in Fig. 2.19

The movement of the mass is the result of an external force $F$ and of the two
forces produced by the spring and by the damper (no gravity force supposed). The
force of the spring is supposed to be equal to $kx$, where $k$ is a constant, and the force
of the damper is *Bdx/dt, B* being a constant. Here, $x$ is the displacement from the
initial equilibrium value. The following figure shows a graph, which describes the
system ($a$ is the acceleration) (Fig. 2.20).

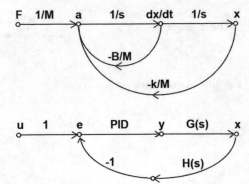

**Fig. 2.20** The signal flow graph for model 2.19

**Fig. 2.21** Signal flow graph of a control system

The corresponding differential equation is

$$(F(t) - kx(t) - Bdx/dt)/M = a.$$

Note that *F(t)* is a *source node* of the graph, i.e., it does not depend on any internal signal of the system.

The graph if Fig. 2.21 represents the dynamics of a feedback control system. The signal *u* is the set point, *e* is the control error, *x* is the controlled physical variable, and *h* is its measured value. The link PID is the controller with proportional–integral–derivative action, *G(s)* is the transfer function of the controlled process, and *H(s)* is the dynamics of the measurement instrument. Using SFGs, the user paints the graph on the screen and then specifies the transfer functions for the links. The simulation software does the rest.

### 2.12.2  Mason's Gain Formula

One of the advantages of signal flow graphs is that the total transfer function of complicated SFGs can be calculated from the following formula of Mason [24]. The version described below works for the SFGs that do not contain non-linear links. Using the formula of Mason, we don't have to manage the differential equations of the model. We must select two nodes as input and output for the transfer function we want to calculate. One of them must be a source node, and the other a non-source node.

Let's define the following terms.

A *trajectory, or path* between two nodes (starting and final) is a succession of consecutive links, without contradicting the direction of arrows and without passing by any node two or more times. This means that the path cannot be a closed loop of links.

A *loop of order one, or simple loop* is defined in a similar way as the path. However, for the simple loop, the starting and final nodes are identical.

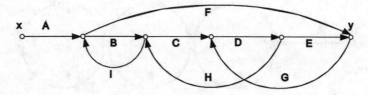

**Fig. 2.22** A signal flow graph, example 1

A *loop of order k>1* is a collection of $k$ simple loops that do not have any common nodes.

The gain of a path, as well as of the simple loop is the product of the gains of the consecutive links that form the path or the loop.

The *determinant* of the graph is defined as follows:

$$\Delta = 1 - \sum_{k=1}^{N_1} M_k^{(1)} + \sum_{k=1}^{N_2} M_k^{(2)} - \sum_{k=1}^{N_3} M_k^{(3)} + \sum_{k=1}^{N_4} M_k^{(4)}, \qquad (2.57)$$

where $M_k^{(i)}$ is the gain of the k-th loop of order $i$, and $N_i$ is the number of loops of order $k$. The determinant of an empty graph is supposed to be equal to one.

The *cofactor* $\Delta_n$ of the path $T_n$ is the determinant of the graph obtained by suppressing all nodes that belong to the path, and all links that enter or go out of them. Below, $T_k$ denotes the $k$-th path that begins in our source node and terminates in the final node.

Finally, the rule of Mason is as follows:

$$G(s) = \frac{\sum_{k=1}^{N} T_k \Delta_k}{\Delta}, \qquad (2.58)$$

**Example 1** Consider the graph of Fig. 2.22

Capital letters are link names and also the corresponding transfer functions. Let calculate the transfer function between nodes $x$ (source) and $y$. There are two paths from $x$ to $y$:

$T_1$ with links $A, B, C, D, E$, and $T_2$ with links $A$ and $F$.

The corresponding cofactors are: $\Delta_1 = 1$, and $\Delta_2 = 1 - CDH$.

The last cofactor was calculated as the determinant of the graph that remains after suppressing the links $A$ and $F$.

The determinant of the whole graph is as follows:

$$\Delta = 1 - BI - CDH - DEG + BIDEG. \qquad (2.59)$$

So, the gain from $x$ to $y$ is

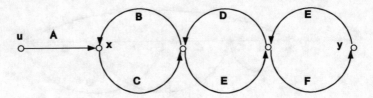

**Fig. 2.23** A signal flow graph, example 2

$$G_{x \to y} = \frac{ABCDE + AF(1 - CDH)}{\Delta}. \tag{2.60}$$

**Example 2** Let calculate the transfer function between nodes $u$ and $x$ of the graph 2.23.

We have only one path $T_1$ from $u$ to $x$, with gain $A$. The cofactor of this path is

$$\Delta_1 = 1 - DE - EF$$

The determinant of the whole graph is

$$\Delta = 1 - BC - DE - EF + BCEF. \tag{2.61}$$

So, the transfer function from $u$ to $x$ is

$$G_{u \to x} = \frac{A(1 - DE - EF)}{1 - BC - DE - EF + BCEF}. \tag{2.62}$$

Show that the transfer function between $u$ and $y$ is as follows:

$$G_{u \to y} = \frac{ACEF}{1 - BC - DE - EF + BCEF}. \tag{2.63}$$

**Example 3** First of all, we should look at the graph and, if possible, simplify it. For example, if we want to calculate the transfer function between nodes **a** and **b** of the graph of Fig. 2.24, then we can ignore the nodes **e f** and **g** with the corresponding links because what happens in this part of the graph, has no influence on the resulting signal at **b**. However, if the output node is **g** or **f**, then we must analyze the entire graph (Fig. 2.25).

A question may arise about the gain between nodes **a** and **b**. The determinant of the whole graph depends on all loops; so, why the gain $G_{a \to b}$ does not depend on the links between nodes **e f** and **g**? Let us calculate the gain, according to Mason's rule. Indeed, the overall determinant is as follows:

**Fig. 2.24** A signal flow graph

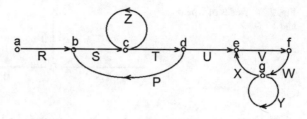

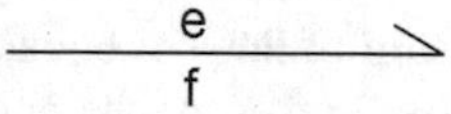

**Fig. 2.25** A bond

$$\Delta = 1 - Z - STP - XVW - Y + ZY + ZXVW + STPXVW + STPY.$$

We have one path from **a** to **b** with gain $T_1 = R$. The corresponding cofactor is $\Delta_1 = 1 - Z - XVW - Y + ZY + ZXVW$. Thus, the gain from **a** to **b** is

$$G_{a \to b} = \frac{R(1 - Z - XVW - -Y + ZY + ZXVW)}{1 - Z - STP - XVW - Y + ZY + ZXVW + STPXVW + STPY} \tag{2.64}$$

But,

$$\Delta_1 = (1 - Z)(1 - XVW - Y)$$

and

$$\Delta = (1 - Z - STP)(1 - XVW - Y).$$

So, the term $(1 - XVW - Y)$ disappears from the formula of $G_{a \to b}$, and finally, we have

$$G_{a \to b} = \frac{R(1 - Z)}{1 - Z - STP}. \tag{2.65}$$

### 2.12.3 Bond Graphs

Bond graph is a widespread tool in the modeling of physical systems. The fact that a bond connects two variables: the effort and the flow, makes this tool the most appropriate for modeling systems with energy flow because the power produced at the bond is the product of these two variables (e.g., voltage and current, force and velocity, liquid pressure and flow, etc.). Again, this method permits to eliminate both mathematics and coding from the simulation task. For a good review on bond graphs consult Cellier [7, 8] and Borutzky and Gawthrop [5, 6] and Gmiterko [16].

A bond graph model is composed by the *nodes or junctions* and the *links named bonds*. A bond is a directed link with a harpoon. The harpoon is placed on the left of the link (related to its direction). The two variables are indicated as follows. The effort is placed on the side of the harpoon and the flow is indicated on the other side.

**Fig. 2.26** Nodes of type 0
and 1

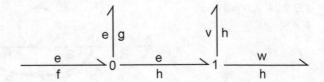

There are several types of nodes in bond graph models. At the node of type 0, the sum of flows is equal to zero, while the efforts of all connected bonds are equal to each other. At the node of type 1, the sum of efforts must be zero and the flows of all corresponding bonds are equal to each other. Thus, we can represent graphically a system that obeys a number of balance equations.

For example, the node equations for the graph of Fig. 2.26 are as follows:

$$f - g - h = 0 \text{ for node of type } 0$$
$$e - v - w = 0 \text{ for node of type } 1$$

The sign of a term in the node equation depends on the direction of the corresponding bond, the outgoing bond having the negative sign of the corresponding variable. Other possible nodes are as follows:

SE node—effort source, e.g., an external force, ideal voltage source, etc.

SF node—flow source, e.g., mandated velocity in a translational system and ideal current source.

R node—dissipative element, e.g., damper or electrical resistance.

C node—capacitance, e.g., a spring or electrical capacitance.

L/I node—Inertia/inductance, e.g., a moving or rotating mass and electrical inductance.

The causality in bond graph diagrams is denoted by a stroke at one of the ends of the bond. This means that the flow variable is evaluated at the end with the stroke and the effort variable at the other side.

For the node of type "0," only one connected bond can have the causality stoke at the side of the node. For the node of type "1," there must be only one connected bond **without** causality stroke at the side of the node.

Figure 2.27 shows the possible combinations of bonds and nodes of type SE, SF, R, L/I, and C, and the implied causalities. All free ends of the bonds can be connected to nodes of type 0 or 1.

The node–bond combinations are

(a) (SE node) Effort source. The effort $e$ is defined at the node.

(b) (SF node) Flow source. The flow $f$ is defined.

(c),(d) (R node) Dissipative bonds. The equations are

$f = e/R$ and $e = f R$, respectively.

(e) (C node) Capacitance. It has the desired causality as shown. The equation is

$de/dt = f/C, C$ being the capacitance.

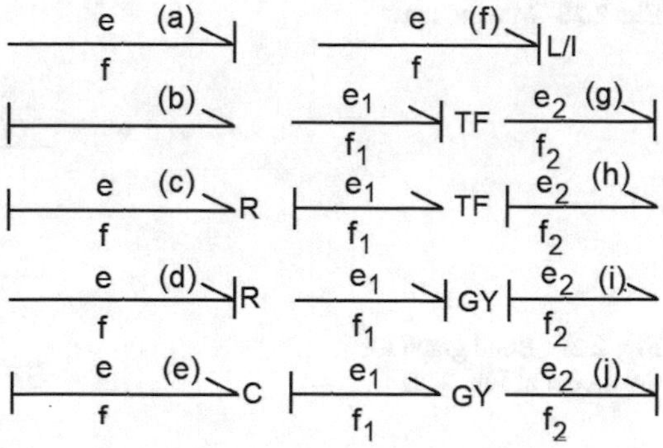

**Fig. 2.27** Bond types

(f) (I/L node) Inertia or Inductance. The equation is
$df/dt = e/L$, $L$ is a constant (mass, inductance, etc.).
(g) (TF bond) Ideal transformer. The equations are
$e_1 = me_2$, $f_2 = mf_1$, $m$ is a constant.
(h) (TF bond) Ideal transformer. The equations are
$e_2 = e_1/m$, $f_1 = f_2/m$, $m$ is a constant.
(i) (GY bond) Ideal "gyrator." The equation is
$e_1 = rf_2$, $e_2 = rf_1$, $r$ is a constant.
(j) (GY bond) Ideal "gyrator." The equation is
$f_2 = e_1/r$, $f_1 = e_2/r$, $r$ is a constant.
Note that the R node–bond combination and the TF and GY bonds have two possible causalities, while the other bonds must have the causalities indicated in the previous figure. The R, TF, and GY causalities are given by the user.

## 2.12.4 Example of Bond Graph

Consider a simple mechanical system shown in Fig. 2.28.
$p = v - w$;
$f$ is the force at point a;
$h$ is the force of the damper;
$h = rp$;
$dg/dt = p/k$;
$r$ is the constant of the damper;
$k$ is the constant of the spring.
In mechanical systems, the efforts are forces and flows are velocities (Fig. 2.29).
The user model needs no causalities to be defined. Good bond graph software must determine causalities automatically, then generate the model equations and corresponding code, and run the simulation. The only situation, when the user must check the causalities is when the software detects a causality conflict. It may result from a physically invalid model. For example, the causality conflict will occur when

**Fig. 2.28** Moving mass

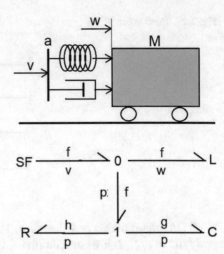

**Fig. 2.29** Bond graph for
the model of Fig. 2.28

you put a voltage source in parallel with a capacitor or a current source in series
with an inductance. Such models imply a possibility of infinite pulses of current or
voltage. This makes it difficult or impossible to calculate the system trajectory, and
the simulation fails.

## 2.12.5   The Causality and DYMOLA

The common approach to the modeling of physical systems is almost always related
to the principle of causality. Note, however, that there are physical objects that do
not necessarily obey the causality rule, or have the cause–effect relation vague or
undefined. For example, the state of an electrical resistor may be described by the
current and the voltage over the resistor, but none of these two variables can be
defined as the cause or an effect. The electric capacitance has this relation better
defined. If we treat the voltage as the cause, then we can calculate the corresponding
current. However, if the voltage (cause) change is a discontinued function of time,
then we must admit an infinite pulse of current at the time instant of discontinuity. If

we treat the current as the cause and the voltage as the result, no such difficulty takes place. Observe that what is essential in bond-graph modeling of physical systems is not the causality, but the modeling of the energy flow.

In the conventional signal-flow or bond graphs, the causality is always well defined. This paradigm repeats in many modeling tasks, but it may be questioned. Celler [9] opinion is that *"Physics, however, is essentially acausal. It is one of the most deep-rooted myths of engineering that algebraic loops in models result from neglected fast dynamics. This myth is based upon the engineers' infatuation with state-space models. Since state-space models are what the engineers know, they believe that this is also how the universe operates. Whenever they encounter an algebraic loop in a model, they introduce a "small capacitor" (a storage element) to break it, and claim that they actually represent the physical realities more faithfully in this way."*

This approach to causality was implemented in the language DYMOLA [9]. This is a general-purpose hierarchical modular modeling software for continuous systems. The main property of DYMOLA is that the software can solve the equation of a device being modeled, whatever the causality for the device could be. This, and the hierarchical model building capacity of DYMOLA, make the language a useful and consistent tool for modelers. For more details on DYMOLA, consult [9].

## 2.13 Models with Distributed Parameters, Partial Differential Equations

We will not discuss the numerical methods for partial differential equations (PDEs). Let us give only some remarks about the difficulties in integrating the equations and the stability of numerical solutions.

### 2.13.1 PDE Solution Algorithms

Partial differential equations (PDE) describe the dynamics of some media, like gas, liquid, or temperature distribution in time and space. To see how complex the problem may be, consider, for example, the following equations of fluid dynamics (Navier–Stokes equations). Here, vector $\mathbf{v} = (u, v, w)$ is the velocity of the medium in a point of three-dimensional space with coordinates $(x, y, z)$, and $t$ represents the time (case of liquid flow)

$$\begin{cases} \rho\left(\dfrac{\partial u}{\partial t} + u\dfrac{\partial u}{\partial x} + v\dfrac{\partial u}{\partial y} + w\dfrac{\partial u}{\partial z}\right) = -\dfrac{\partial p}{\partial x} + \mu\nabla^2 u + \rho f_x \\[3mm] \rho\left(\dfrac{\partial v}{\partial t} + u\dfrac{\partial v}{\partial x} + v\dfrac{\partial v}{\partial y} + w\dfrac{\partial u}{\partial z}\right) = -\dfrac{\partial p}{\partial y} + \mu\nabla^2 v + \rho f_y \\[3mm] \rho\left(\dfrac{\partial w}{\partial t} + u\dfrac{\partial w}{\partial x} + v\dfrac{\partial w}{\partial y} + w\dfrac{\partial w}{\partial z}\right) = -\dfrac{\partial p}{\partial z} + \mu\nabla^2 w + \rho f_z \\[3mm] \dfrac{\partial u}{\partial x} + \dfrac{\partial v}{\partial y} + \dfrac{\partial w}{\partial z} = 0, \\[3mm] where\ \nabla^2 = \left(\dfrac{\partial^2}{\partial x^2}, \dfrac{\partial^2}{\partial y^2}, \dfrac{\partial^2}{\partial z^2}\right) \end{cases} \quad (2.66)$$

Here, $\rho$ is the liquid density, $\mu$ is the viscosity, $p$ is the pressure, and $f$ represented the eventual external force.

The above set of the Navier–Stokes equations is difficult to solve, mainly because of its non-linearity and the number of equations. We have three velocity variables and the pressure. Here, the density is supposed constant and we do not deal with the temperature distribution. This means that, after space discretization, we must solve a set of 4N non-linear equations, where N is the number of the grid points, normally between 50,000 and 1,000,000, which gives up to 4,000,000 equations. These difficulties are related to the lack of compressibility of the liquid.

In the gas case, we have (in vectorial form):

$$\begin{cases} \dfrac{\partial p}{\partial t} + \nabla \cdot (\rho\mathbf{v}) = 0 \\[3mm] \rho\dfrac{\partial v}{\partial t} + \rho\mathbf{v} \cdot \nabla)\mathbf{v} = \nabla\mathbf{p} \\[3mm] \rho\dfrac{\partial e}{\partial t} + \rho(v \cdot \nabla)e = -p\nabla \cdot v + \nabla \cdot (k\nabla T) \\[3mm] where\ \nabla = \left(\dfrac{\partial}{\partial x}, \dfrac{\partial}{\partial y}, \dfrac{\partial}{\partial z}\right) \end{cases} \quad (2.67)$$

Here, $e$ is the gas internal energy, $\rho$ is the gas density,
$T$ is the temperature, $dT = de/C$, $C$—specific heat.

There is a huge literature that describes a great number of numerical methods for the PDEs. In the book of Ames [1] we can find a good overview of these methods. While solving PDEs, the main difficulties are the computer time and the stability of the numerical method. Most methods use a certain discretization scheme, both

**Fig. 2.30** Differentiation
schemes

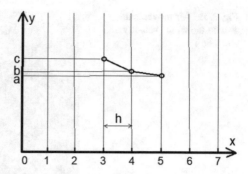

in time and space. The problems arise when we decrease the discretization steps to achieve desired time-space resolution. Observe that the PDEs manage time and space differentiation, given by differential operators. The problem is that these *operators are usually unbounded* on the unit sphere in the space of functions. This is the main cause of *numerical instability*.

For example, the equations (2.67) can be rearranged in such a way that we have the time derivative at the left hand and a differential operator on the right. Consider another very simple example. Let $y$ be the dependent variable which changes in time depend on its derivative with respect to space variable $x$ (see Fig. 2.30). To calculate the increment of $y$ over a consecutive time-step, we must execute the differentiation operation over the space-step of variable $x$ (see Fig. 2.30). So, we can use a finite-step differencing scheme. The simplest are forward- and backward-differencing, that provide the value of the derivative, as follows.

$$d_1 = (y_{n+1} - y_n)/h \text{ and } d_2 = (y_n - y_{n-1})/h, \text{ where } y_n = y(x_n).$$

The problem is, which one of the values $d_1$ and $d_2$ should be used? A logical solution seems to be the use of the average $(d_1 + d_2)/2$. This average provides the value

$$d_n = (y_{n+1} - y_{n-1})/(2h) \tag{2.68}$$

This is called Forward Time-Centered Space (FCTS) scheme. Here, $d_n$ is the estimate of the derivative at $x_n$. Consequently, we have $y_n t + \delta t = y_n(t) + h d_n$.

Now, observe that the increment of $y_n$ depends on $y_{n+1}$ and $y_{n-1}$ and does not depend on $y_n$. Thus, the solution for $y$ in the even space-steps depends on the values at odd steps and vice versa. This may result in solutions where $y$ at even and odd space points are quite different. This fact shows how useless is the FTCS differencing scheme. Moreover, it is known that integration with this differencing scheme is *always unstable*.

The common remedy to the FTCS instability is the LAX method (see Lax [21]). In this method, we replace Eq. (2.68) with the following.

**Fig. 2.31** Air movement
around the wing (velocity
elements)

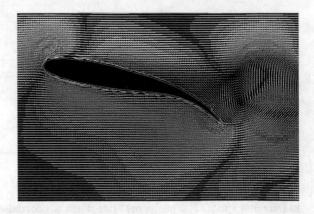

$$d = (y_{n+1} - y_{n-1})/(2h) - \frac{v\Delta t}{2h}(d_{n+1} - d_{n-1}). \tag{2.69}$$

The last term of (2.69) stabilizes the scheme. Here, $v$ is a constant, sometimes called wave propagation speed, and $\Delta t$ is the time-step.

The LAX method, however, is not any "magic" remedy for the PDE solution. In general, the use of (2.69) is equivalent to adding some artificial viscosity or dissipation, to the solution. This results in stable solutions but by adding a dissipation to the model we, in fact, solve an altered problem.

There are a lot of other methods that make the differencing schemes stable. Unfortunately, these methods almost always alternate the original model. Another inconvenience is that the increase of the time–space resolution results in a rapid increment of the computing time. Better results are obtained by the *finite element method*, shortly described in the following section (Figs. 2.31 and 2.32).

### 2.13.2   Finite Element Model

The main difference between finite step differencing algorithms and the *finite element method* is that the space discretization is not uniform. A *mesh* of small *finite elements* is defined. The elements need not be cubes. They are rather volume elements with different sizes and shapes that fill the volume under consideration. Depending on the physical properties of the medium (gas, liquid, heat, electromagnetic field, etc.), the algebraic equation for the finite element is defined. This leads to the problem of solution to a number of algebraic equations, solved by the methods of the calculus of variations and optimization.

The main software tool for finite element method is the ANSYS package, equipped with a sophisticated numerical algorithm, graphical user interface, and 3D results display. We will not discuss here the details of the finite element method because

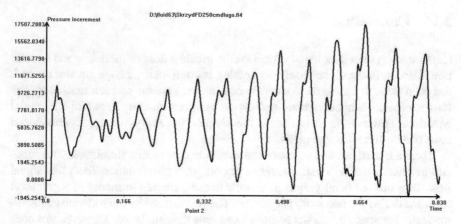

**Fig. 2.32** The low-frequency oscillations during 8 seconds of flight

the topic of this book is not numerical analysis but rather general concepts of model building. For more detail, consult [3, 10, 27].

### 2.13.3   Example: Jet Takeoff Vibrations

In many occasions, passengers complain about the strong vibrations of the fuselage during the plane lift-off. This makes the takeoff unpleasant. One of the causes of vibrations may be the air turbulence below the wing. This can be simulated by solving the Navier–Stokes equations. Such simulations have been described in [28]. Let's show here only the most relevant result. Figure 2.31 shows the velocity field around the wing just after takeoff. This is a two-dimensional projection of the three-dimensional simulation.

On the animated simulation screen, it can be seen that around the wing big vibrations and turbulences appear that generate the vibrations in air pressure. This results in strong vibrations of the whole fuselage and may even be dangerous for the struture of the fuselage. The model and simulation experiments have been done using the Fluids6 simulation program, consult [28] for more detail.

The simulations of this, and similar models show that the fluid flow hardly reaches a steady state. Such a state may satisfy the fluid equations, but in reality, never takes place. So, many of the results provided by the fluid dynamics software may be correct from the mathematical point of view, but useless and confusing in design tasks because the velocity field is always in movement. The above modeling and simulations are important for safety issues. Figure 2.32 shows the low-frequency pressure oscillations below the wing.

## 2.14 Conclusion

Continuous system simulation is perhaps the greatest field of modeling and simulation. Though in the digital computer nothing is continuous, a huge amount of work has been done to enable us to simulate continuous systems on such machines. The most used models of continuous systems are based on ordinary or partial differential equations. However, as mentioned in this chapter, not everything that is continuous must obey differential equations.

There are methods and software that facilitate continuous simulation. The simulationist needs not be a mathematician or computer programmer. Tools like signal flow diagrams, or bond graphs permit the user to create the model in a graphical form and define the necessary parameters. The rest is done by the software that automatically generates the model equations and runs the simulation. However, this does not mean that everybody can simulate. To create a model of any form, the person must have some knowledge on dynamic systems and on the purpose and limitations of the modeling and simulation task. We must remember that creating a differential equation model we must also be aware of the existence, uniqueness, and stability of the solutions.

## 2.15 Questions and Answers

**Question 2.1** Give an example of *concentrated parameter* system and *distributed parameter* system.

**Question 2.2** How the analog computer works?

**Question 2.3** Explain the concept of the *state* of a dynamic system.

**Question 2.4** Give examples of the following systems.
1. Finite automata
2. Infinite automata
3. Concentrated parameter system
4. Distributed parameter system

**Question 2.5** Convert the following equation:

$$\frac{dx}{dt} + \frac{d^3x}{dt^3}(1 + x(t)) - 2 = 0$$

into a system of equations of the first order (canonical form)

**Question 2.6** Given the model of a dynamic system, as follows:

$$\frac{d^2x}{dt^2} + 3\frac{dx}{dt} + x(t) = u(t)$$

find the transfer function, supposing that $u(t)$ is the input signal, and $x(t)$ is the model output.

**Question 2.7** The transfer function depends on:(select)
1. The model time
2. The input signal $u(t)$ or $u(s)$
3. The output signal $x(t)$ or $x(s)$
4. Both signals
5. None of the above

**Question 2.8** What is the transfer function of time delay element $x(t) = u(t - \tau)$? Provide a proof.

**Question 2.9** Calculate the transfer function $y/p$ of the automatic control circuit shown in Fig. 2.33. The controller is supposed to be of PI (proportional-integral) type.

The transfer function of the PI controller is

$$C(s) = K\left(1 + \frac{1}{T_i s}\right)$$

**Question 2.10** Consider a control circuit of Fig. 2.33, with proportional controller, gain $K > 0$.. Using the Routh–Hurwitz criterion, find a range of the gain K, for which the circuit is stable.

**Question 2.11** Given a transfer function $G(s)$, how can we obtain the frequency response of the corresponding dynamic system?

**Question 2.12** What is a *sampled data system*?

**Question 2.13** What is the relation between the variable "z" in Z-transform to the variable "s" of Laplace?

**Question 2.14** Let $y_k$ be a signal delayed by $r$ periods $T$ with respect to $x_k$: $y_k = x_{k-r}$ What is the relation between the corresponding Z-transforms?

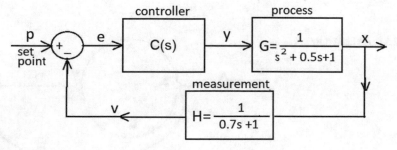

**Fig. 2.33** A feedback control circuit

**Question 2.15** A continuous dynamic system has the following transfer function:

$$G(s) = \frac{1}{s^2 + 5s + 6}.$$
(2.70)

Calculate the sampled data Z-transform version of the transfer function $G(z)$.

**Question 2.16** The spectrum of an analog signal contains frequencies of 0–25Hz. What is the minimum frequency of sampling that permits to completely restore the original signal from the sampled data?

**Question 2.17** What is BIBO stability?

**Question 2.18** Give an example of orbitally stable feedback control system.

**Question 2.19** What is the stability condition for the linear sampled data system, in terms of Z-transfer function?

**Question 2.20** What assumptions about the feedback circuit must be satisfied, to apply the method of *describing function*?

**Question 2.21** In *signal flow graphs*, what represent nodes and links?

**Question 2.22** What is the main field of application of *signal flow graphs*?

**Question 2.23** Calculate the transfer function for the flow diagram of Fig. 2.34, for nodes u→x, and u→z.

**Question 2.24** Construct the bond graph for the electric circuit of Fig. 2.35. Put the causality strokes.

**Question 2.25** What is wrong in this bond graph? (Fig. 2.36)

**Question 2.26** What is the main concept of causality treatment and closed algebraic loops in DYMOLA?

**Question 2.27** What kind of model represents the Navier-Stokes equation? How to classify this kind of dynamic system?

**Fig. 2.34** A signal flow graph

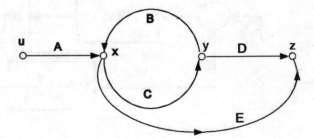

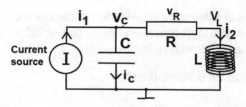

**Fig. 2.35** An electric circuit

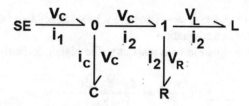

**Fig. 2.36** A bond graph

**Question 2.28** What is the Forward Time-Centered Space (FTCS) algorithm? What are the disadvantage of FTCS?

**Question 2.29** What is the it Finite Element Method?

**Answers**

**Answer 2.1** Concentrated parameter systems:
  1. A rigid body moved by a spring and damper. The spring and the dumper are supposed "ideal"; they have no mass.
  2. RLC electric circuit with ideal elements R, L, and C.

Distributed parameter system: the temperature control in a piece of metal, with dynamics described by the equation of temperature propagation in 3D body.

**Answer 2.2** An analog computer consists of several (maybe 100 or more) *operational amplifiers*. The operational amplifier (OPAM) supports signals of the great frequency spectrum, including frequency zero (DC component), and its gain is big, theoretically infinite. Combining OPAMS with capacitors and resistances, we get integrators and finite gain amplifiers. Some non-linear elements are also available. Then connecting adequately these devices, we obtain a circuit that satisfies a given differential equation. Running the ciruit, we get the solution to the equation.

**Answer 2.3** The system state is a subset of model descriptive variables. We require that given the system state at a moment $s$ and the model inputs over the interval [s,t] we can calculate the state at t (and overall interval [s,t])

**Answer 2.4**
  1. Electric switch

2. Model of a waiting line without limitation of the line lenght
3. Electric circuit with ideal capacitors, resistors, and inductances
4. 3D fluid flow in a duct with obstacles.

**Answer 2.5**  The original equation is as follows

$$\frac{dx}{dt} + \frac{d^3x}{dt^3}(1 + x(t)) - 2 = 0.$$

Here, we suppose that $x(t) \neq -1$ for the whole trajectory $x(t)$ over the interval where we want to solve the equation.

First, we solve the equation with respect to the higher derivative of $x$.

$$\frac{d^3x}{dt^3} = \frac{2 - dx/dt}{1 - x(t)}. \tag{2.71}$$

Now, introduce the following notation:

$$x_1 = x, \quad x_2 = \frac{dx}{dt}, \quad x_3 = \frac{d^2x}{dt^2}.$$

So, we have

$$\frac{dx_1}{dt} = x_2,$$

$$\frac{dx_2}{dt} = x_3,$$

$$\frac{dx_3}{dt} = \frac{2 - dx_1/dt}{1 - x_1(t)},$$

where the derivative of $x_3$ is given by Eq. 2.71.

The above three equations of order one are equivalent to the original equation of order three (canonical form).

**Answer 2.6**  The original equation is as follows:

$$\frac{d^2x}{dt^2} + 3\frac{dx}{dt} + x(t) = u(t).$$

Applying the Laplace transform to both sides of this equation, we get

$$x(s)s^2 + 3x(s)s + x(s) = u(s).$$

So, the transfer function is

$$G(s) = \frac{x(s)}{u(s)} = \frac{1}{s^2 + 3s + 1}$$

**Answer 2.7** Option 5.

The transfer function is a function the complex variable $s$. The function is the property of the model, and does not depend on particular signals $x(t)$, $u(t)$, $x(s)$, or $u(s)$.

**Answer 2.8** In terms of the L-transform, we have $x(s) = e^{-\tau s} u(s)$, so, the transfer function is

$$G(s) = e^{-\tau s}.$$

**Answer 2.9** From the block diagram 2.33, we have

$$\begin{cases} e = p - v = p - Hx \\ x = CGe \\ x = CG(p - Hx) \end{cases} \tag{2.72}$$

Finally, we have

$$x + CGHx = CGp$$

and the transfer function $p \to x$ is as follows:

$$G_{p \to x}(s) = \frac{C(s)G(s)}{1 + C(s)G(s)H(s)},$$

where $C(s) = K(1 + 1/(T_i s))$.

**Answer 2.10** Using the transfer function derived in answer 2.9, we have

$$G_{p \to x}(s) = \frac{KG(s)}{1 + KG(s)H(s)} =$$

$$\frac{K \frac{1}{s^2 + 0.5s + 1}}{1 + K \frac{1}{s^2 + 0.5s + 1} \frac{1}{0.7s + 1}}.$$

Now, multiply the nominator and denominator of the above equation by $(s^2 + 0.5s + 1)(0.7s + 1)$:

$$G_{p \to x}(s) = \frac{K(0.7s + 1)}{(s^2 + 0.5s + 1)(0.7s + 1) + K} =$$

$$\frac{K(0.7s + 1)}{0.7s^3 + 1.35s^2 + 1.2s + 1 + K}.$$

Now, we construct the Routh–Hurwitz array for $N(s) = 0.7s^3 + 1.35s^2 + 1.2s + 1 + K$

| 3 | 0.7 | 1.2 |
|---|-----|-----|
| 2 | 1.35 | $(1 + K)$ |
| 1 | $0.92 - 0.7K$ | – |
| 0 | $1 + K$ | – |

So, the stability condition follows from the requirement that the first element for row number 1 be non-negative:

$0.92 - 0.7K \geq 0$. This means that $K \leq 1.31429$, approximately. If this element is equal to zero, then the circuit is at the stability limit.

**Answer 2.11** Replace the complex variable $s$ with $j\omega$: $G(s) \to G(j\omega)$.

**Answer 2.12** In a sampled data system there is one or more elements (samplers) that react to the input signal only at discrete time instants, with period $T : t = 0, T, 2T, 3T, 4T \ldots$

**Answer 2.13** $z = e^{sT}$

**Answer 2.14** $\mathcal{Z}\{(y_k\} = z^{-r}\mathcal{Z}\{(x_k\}$

**Answer 2.15** 1. Using the difference equations:

According to Eq. (2.70), the differential equation of the model is

$$\frac{d^2x}{dt^2} + 5\frac{dx}{dt} + 6 = u(t), \tag{2.73}$$

where $u(t)$ is the input signal. For discrete time with period T we obtain the following approximation in the form of a difference equation:

$$\frac{\Delta^2 x_k}{T^2} + 5\frac{\Delta x_k}{T} + 6x_k = u_k, \tag{2.74}$$

where $\Delta x_k = x_k - x_{k-1}, \quad \Delta^2 x_k = x_k - 2x_{k-1} + x_{k-2}$.

So, in terms of Z-transform, we have

$$\frac{(x(z) - 2x(z)z^{-1} + x(z)z^{-2})}{T^2} + 5\frac{(x(z) - x(z)z^{-1})}{T} + 6x(z) = u(z), \tag{2.75}$$

$$G(z) = \frac{x(z)}{u(z)} = \frac{1}{(1 - 2z^{-1} + z^{-2})/T^2 + 5(1 - z^{-1})/T + 6}. \tag{2.76}$$

**Answer 2.16** 50 Hz.

**Answer 2.17** BIBO means *Bounded Input, Bounded Output*. It is required that if we apply a bounded external excitation to the system, the corresponding response be bounded also.

**Answer 2.18**   A circuit for temperature control with a two-point (on/off) temperature controller.

**Answer 2.19**   If the transfer function has the form

$$G(z) = \frac{M(z)}{N(z)},$$

where M and M are polynomials, then all the roots of the characteristic equation $N(z) = 0$ must be included in the unit circle around the origin, on the complex plane.

**Answer 2.20**   To use the *describing function* method, the following conditions must be satisfied.

1. The output signal from the non-linear element must have the converging Fourier series representation.

2. The linear part of the feedback circuit must be a low-pass filter.

**Answer 2.21**   The nodes are signals (model variables) and the links are transfer functions or other operators.

**Answer 2.22**   *Signal flow graphs* can be used to represent the dynamics of control systems, instrumentation, and other general dynamic systems.

**Answer 2.23**   The determinant of the whole graph is $\Delta = 1 - BC - CDE$.

1. Transfer function $u \to x$:

There is one trajectory $T_1 = A$. Cofactor of this trajectory is equal to 1.

$$\text{Transfer function } u \to x = \frac{A}{1 - BC - CDE}.$$

2. Transfer function $u \to z$:

There are two trajectories: $T_1 = ACD$ and $T_2 = AE$. Cofactors of both trajectories are equal to 1.

$$\text{Transfer function } u \to z = \frac{ACD + AE}{1 - BC - BDE}.$$

**Answer 2.24**   The bond graph is as follows (Fig. 2.37)

**Answer 2.25**   Try to put the correct causality strokes at the graph. You will see that there is a causality conflict in this graph.

This conflict is the result of an invalid model. For example, it may be an electric circuit like that of Fig. 2.35, where instead of the current source there is a voltage source. You cannot connect in parallel a voltage source with a capacitor.

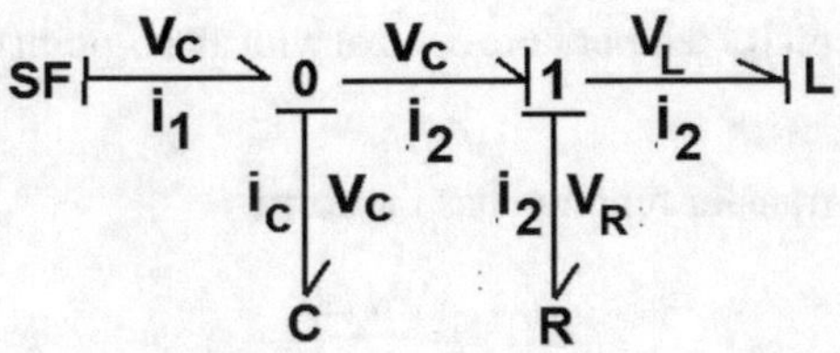

**Fig. 2.37** A bond graph for circuit 2.35

**Answer 2.26** In DYMOLA, the software can solve the equations of the model, whatever the causalities for the model component could be. the software also solves the closed algebraic loops, if any.

**Answer 2.27** The Navier–Stokes equation describes the fluid dynamics, gas or liquid. It is a non-linear system of partial differential equations. The model can be classifies as the *distributed parameter system*.

**Answer 2.28** FTCS scheme is one of the numerical methods for partial differential equations. The main disadvantage of FTCS is that it is always unstable.

**Answer 2.29** In the *finite element model*, the space discretization is not uniform. The method uses a mesh of small *finite spatial elements* with different sizes, for spatial discretization.

# References

1. Ames WF (1983) Numerical methods for partial differential equations, 2nd edn. Academic Press, New York
2. Bader G, Deufthard P (1983) A semi-implicit mid-point rule for stiff systems of ordinary differential equations. Numerische Mathematik 41:373–398
3. Bathe KJ (1976) Numerical methods in finite element analysis. Prentice-Hall. ISBN/ISSN. https://doi.org/10.1080/10.0136271901
4. Bhatia NP, SzegoG P (1970) Stability theory of dynamical systems. Springer-Verlag, Berlin
5. Borutzky W (2011) Bond graph modelling of engineering systems: Theory. Springer, Applications and Software. ISSN. 10.1441993673
6. Borutzky W, Gawthrop P (2006) Bond graph modelling. Math Comput Model Dyn Syst 12(2–3):103–105. https://doi.org/10.1080/13873950500069078
7. Cellier F (1992) Hierarchical non-linear bond graphs: a unified methodology for modeling complex physical systems. Simulation 55(4):230–248. https://doi.org/10.1177/003754979205800404
8. Cellier FE, Greifeneder J (1991) Continuous system modeling. Springer. ISBN/ISSN 978-1-4757-3922-0
9. Cellier FE, Elmquist H (1995) Automated formula manipulation supports object-oriented continuous system simulation. IEEE Cont Syst 13(2):28–38
10. Chaskalovic J (2008) Finite elements methods for engineering sciences. Springer. ISBN/ISSN 978-3-540-76343-7

11. Chen G (2004) Encyclopedia of RF and microwave engineering. Wiley, New York, pp 4881–4896
12. Collatz, Funktionalanalysis und Numerische Mathematik. Springer, ISBN/ISSN ISBN-10:3642950299
13. Dahlquist G, Bjorck A (2012) Numerical methods. Dover Publications, ISBN/ISSN, p 9780486139463
14. Dahlquist G, Bjork A (1974) Numerical methods. Prentice Hall
15. Faulkner EA (1969) Introduction to the theory of linear systems. Chapman & Hall. ISBN 0-412-09400-2
16. Gmiterko A, Lipták T, (1969) The usage of bond graphs methodology for mechanical systems designing. Appl Mech Mater 816(816):349–356. https://doi.org/10.4028/www.scientific.net/amm.816.349
17. Hahn H (1967) Stability of motion. Springer-Verlag, Berlin
18. Hinrichsen D, Pritchard AJ (2005) Mathematical systems theory I-modelling, state space analysis. Springer Verlag, Stability and Robustness. https://doi.org/10.1080/9783540441250
19. Ifeachor EC, Jervis BW (1993) Digital signal processing. ISBN, Addison-Wiley, p 029154413X
20. Krylov NM, Bogoliubov N (1943) Introduction to nonlinear mechanics. Princeton Univ. Press, Princeton, US. 0691079854. Archived from the original on 2013-06-20
21. Lax PD, Wendroff B (1960) Systems of conservation laws. Commun Pure Appl Math 13(2):217–237. https://doi.org/10.1002/cpa.3160130205
22. Marks RJ (1991) Introduction to shannon sampling and interpolation theory. Springer-Verlag
23. Martynyuk AA (ed) (2003) Advances in stability theory at the end of the 20th century. Taylor & Francis, London
24. Mason SJ (1956). Feedback theory—further properties of signal flow graphs. Proceed IRE: 920–926
25. McNamara C (2006) Field guide to consulting and organizational development. Authenticity Consulting, LLC. 10:1933719206
26. Page SE (2018) The model thinker. Hachette Book Group, New York. 978-0-465-009462-2
27. Matsson J (2020) An introduction to ANSYS Fluent 2020. SDC Publications, ISBN/ISSN, p 1630573965
28. Raczynski S (2016) Takeoff vibrations of a jetliner: simulating possible cause. Conference paper: annual simulation symposium, the society for modeling and simulation (the society for modeling and simulation, eds.) , Pasadena CA, ISBN/ISSN 1-56555-359-4, 2016
29. Sagawa JK, Nagano MS (2015) Applying bond graphs for modelling the manufacturing dynamics. Elsevier. https://doi.org/10.1016/j.ifacol.2015.06.390
30. Tavangarian D, Waldschmidt K (1980) Signal flow graphs for network simulation. Simulation 34(3):79–92. https://doi.org/10.1177/003754978003400308
31. Willems JL (1970) Stability theory of dynamical systems. UK, Nelson, p 0177810068

# Chapter 3
# Differential Inclusions, Uncertainty, and Functional Sensitivity

## 3.1 Introduction, Some Definitions

In any field of scientific research, we should look for improved or new methods. As stated in the previous chapters, the division of modeling and simulation into discrete and continuous simulation may be useful, but from the methodological point of view, it is somewhat artificial. As for the continuous models, the ordinary and partial differential equations are not the only possible modeling tools. Also, the problem of uncertainty in models may have a new insight, using an alternative model type, like *differential inclusions*.

The central issue in practical applications of differential inclusions is the problem of the determination of reachable sets. In this chapter, we will discuss the algorithm of the *differential inclusion solver* and its application to the general task of modeling and in the problem of uncertainty treatment. Also, the concept of the *functional sensitivity* is discussed.

For a comprehensive overview of the topic, consult Aubin and Cellina [2, 4]. Here, let us recall the main concepts.

For the early publications on DIs date from the 1930s, see papers of Marchaud [21], Zaremba [48], Wazewski [43–46], Plis [28], and Turowicz [39, 40]. We delimit the definitions to the real Euclidean n-dimensional space $R^n$. Some generalizations of differential inclusions and properties of the reachable sets in more abstract, Banach spaces can be found in the Journal of Mathematical Analysis and Applications, see Raczynski [33, 34].

We assume that the reader is familiar with the terms *almost everywhere, Lipshitz condition, absolutely continuous function, and set-to-point distance*.

Let $X$ and $Y$ be two non-empty subsets of a metric space. The *Hausdorff distance* between $X$ and $Y$ is defined as follows:

$$d_H(X, Y) = max \left\{ \sup_{x \in X} \inf_{y \in Y} d(x, y), \sup_{y \in Y} \inf_{x \in X} d(x, y) \right\}, \qquad (3.1)$$

© The Author(s), under exclusive license to Springer Nature Switzerland AG 2022
S. Raczynski, *Models for Research and Understanding*, Simulation Foundations, Methods and Applications, https://doi.org/10.1007/978-3-031-11926-2_3

where *sup* represents the least upper bound and $d(*,*)$ is the distance between two points. The Hausdorff distance permits to use the concept of continuity of set-valued functions. We say that a mapping from the real line to the space of closed subsets of $R^n$ is *continuous in Hausdorff sense* if it is continuous in the sense of the Hausdorff distance (in the topology induced by the Hausdorff distance).

Suppose that $f$ is a real-valued function $f : R^n \to R$, $x_0 \in R^n$, and $f$ has a finite value at $x_o$. The function $f$ is *lower semi-continuous* at $x_0$ if for every $\varepsilon > 0$ there exists a neighborhood $U$ of $x_0$ such that

$$f(x) \le f(x_0) - \varepsilon, \ \forall x \in U. \tag{3.2}$$

Function $f$ is *upper semi-continuous* at $x_0$ if for every $\varepsilon > 0$ there exists a neighborhood $U$ of $x_0$ such that

$$f(x) \ge f(x_0) + \varepsilon, \ \forall x \in U. \tag{3.3}$$

A function is upper or lower semi-continuous over an interval if the above condition holds for all $x_0$ in the interval under consideration. Figure 3.1 shows an example of a lower semi-continuous function. Note that the function value at $x_o$ is defined as shown by the black dot. Roughly speaking, the function cannot have "jumps" to lower values at any point where it is defined.

The continuity can be defined also for set-valued functions, using the metric induced by the Hausdorff set-to-set distance. As for the lower semi-continuity, the following definition is used.

Let $F$ be a mapping from $R^n$ to subsets of $R^n$. We say that $F$ is lower semi-continuous (l.s.c.) at $x_0 \in R^n$ if and only if for any open set $V \subset R^n$, such that $F(x_0) \cap V \ne \emptyset$, there exists a neighborhood $U \subset R^n$ of $x_0$ such that

$$\forall x \in U : F(x) \cap V \ne \emptyset. \tag{3.4}$$

The mapping $F$ is said to be l.s.c. on $R^n$ if $F$ is l.s.c. for every $x \in R^n$.

The lower semi-continuity is an important property of certain set-valued functions associated with the function that is used as the right-hand side of the differential inclusion, described further on.

**Fig. 3.1** A lower
semi-continuous function

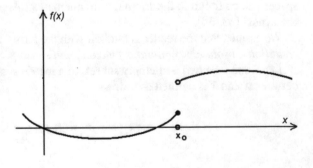

## 3.2 Differential Inclusions

Consider a mapping $F : R^n \times I \to P(R^n)$, where $P(X)$ denotes the power set, i.e., set of all subsets of the space $X$, $I$ is an interval, $I \subset R$. A selection (or selector) of $F$ over $I$ is a function $z(t)$, such that $z(t) \in F(x, t) \forall t \in I$. The existence of selectors is a consequence of the known Axiom of Choice, see Halmos [18].

Some facts related to selections are quite interesting and may contradict our intuition. For example, this is not true that a continuous field should have a continuous selector. Aubin and Cellina [4] show an example of such field. On the other hand, a discontinuous field may have continuous selections. For example, let the value of $F : [0, 2] \to R^2$ be a filled rectangle with vertices
    $(-1, 1), (1, 1), (1, -1)$, and $(-1, -1)$ for $t \leq 1$
and a filled rectangle
    $(-0.5, 0.5), (0.5, 0.5), (0.5, -0.5)$, and $(-0.5, -0.5)$ for $t > 1$.
The field is discontinuous at $t = 1$. However, it has a continuous selection $z(t) \equiv 0$.

**Differential Inclusion** (DI) is defined by the following statement:

$$\frac{dx}{dt} \in F(x, t), \quad x(0) \in X_0, \tag{3.5}$$

where $t$ is a real variable (representing the time in this chapter), $x$ is a function of time, $x(t) \in R^n$, $F$ is a mapping from $R^n \times R$ to subsets of $R^n$, and $X_0 \in R^n$ is the initial set. $R^n$ is the real n-dimensional Euclidean space. In the following, $R$ denotes $R^1$. $F$ is also called the set of *admissible directions*. In the original Wazewski works on DIs, the field of directions in $R^n$, defined by the multi-valued function $F(x, t)$ of (3.5) is called *orientor field*.

We will call the function $x(t)$ a *trajectory* of the DI, if it satisfies (3.5) over the interval under consideration. The trajectory must be absolutely continuous and almost everywhere differentiable function.

## 3.3 Reachable Set

Recall that the graph of a function $f(t)$ is the set of all ordered pairs $(t, f(t))$. Let $X_0$ be a closed and connected subset of $R^n$, $I$ denotes an interval $[t_0, t_1]$, $x(t) \in R^n$ is the model state vector, and $F : R^n \times R \to P(R^n)$ is a set-valued function, where $P(X)$ denotes the power set, i.e., set of all subsets of a space $X$.

The *reachable or attainable set* (RS) of (3.5) is defined as the union of the graphs of all trajectories of (3.5). The term *emission zone* has also been used in early works. In many works on the DIs, the mapping $F$ is called a *field of permissible directions*, and a trajectory of the DI is also called a trajectory of the field $F$.

Let us comment on the term "solution to the DI." It is commonly understood that a trajectory of the DI is it's solution. Observe however that a DI normally has an infinite number of trajectories. Thus, the trajectory cannot be just called "the solution." Our point is that THE solution to a differential inclusion is given by its reachable set. Consider a sequence of DIs with shrinking right-hand side that, in limit, degenerates to a single-valued function. The corresponding sequence of reachable sets tends to the graph of the solution of the resulting differential equation. This is an argument to call the reachable set the solution to the DI. However, to avoid ambiguity and conflict of terms, the term "solution to a DI" will not be used in the following sections. Instead, we will discuss trajectories and reachable or attainable sets.

An absolutely continuous function $x(t)$ is called a quasitrajectory of the DI (3.5) over an interval $I$ with initial condition $x_0$, if a sequence of absolutely continuous functions $\{x_i\}$ exists such that

$$\begin{cases} (i) \;\; x_i(t) \to x(t) \;\; \forall t \in I = [t_0, t_1] \\ (ii) \;\; d(x'(t), F(x(t), t)) \to 0 \;\; a.e.\,on\; I \\ (iii) \;\; x_i' \; are\; equibounded\; on\; I \\ (iv) \;\; x(0) = x_0, \end{cases} \qquad (3.6)$$

where the prime mark stands for time differentiation.

Recall that a sequence of functions $x_k : [t_0, t_1] \to R^n$, $k = 1, 2, 3 \ldots$, $t \in [t, t_1]$, $t_0 < t_1$ is said to be equibounded if such $M$ exists that

$$|x_k(t)| \leq M \; \forall k = 1, 2, 3, \ldots \; and\; t \in [t_0, t_1]. \qquad (3.7)$$

A function $x(t)$ is called a *strong quasitrajectory* of the field $F$, if there exists a sequence $\{x_i(t)\}$ of trajectories of $F$, such that $x_i(t) \to x(t)$ in $[t_0, t_1]$.

Turowicz [40] has given some sufficient conditions for a quasitrajectory to be strong. Let us notice that the notion of strong quasitrajectory ("sliding regime") was introduced independently and earlier by Filippov [13] under stronger hypotheses. The set $E = conv(F)$ is defined as the smallest convex and closed hull of the set $F$, and the set $Q = tend(F)$ is the smallest closet subset of $E$ that has the same convex hull, $conv(Q) = conv(F)$, see Fig. 3.2. Note that the *tendor set Q* of Fig. 3.2 contains the curved sections a and b and the point c.

A very useful property of quasitrajectories of the fields $F$, $E$, and $Q$ was found by Wazewski. He pointed out that, if the field $F$ is continuous in the Hausdorff sense, then the field $E$ is also continuous and the field $Q$ is lower semi-continuous. The most important property of these fields is that, under some additional regularity assumptions, **the fields F, E, and Q have the same quasitrajectories**. Moreover, the Filippov–Wazewski theorem states that, if $F$ satisfies the Lipschitz condition, then for each quasitrajectory, a sequence of trajectories exists that converges to this trajectory. Consequently, the three fields have the same reachable sets with accuracy to their closures (reachable sets of the fields $F$ and $Q$ need not be closed).

This means that in any neighborhood of a trajectory of the field $Q$, a trajectory of the field $F$ exists. This also means that, in many cases of control systems, the

**Fig. 3.2** Example of sets F,
E, and Q

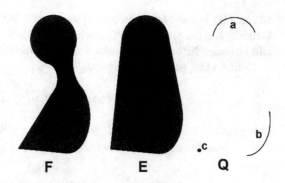

"tendor" or "bang-bang" type of control (field $Q$) can be used without restricting the
system reachable set. See also Raczynski [33, 34].

Let $T = (-\infty, \infty)$, $W = R^n \times T$. Consider the following hypothesis:

**Hypothesis H(F):** for each $(x, t) \in W$, $F(x, t) \in comp(R^n)$, $F(x, t)$ is
bounded, and continuous on $W$.

Wazewski [43] pointed out that under Hypothesis H(F), we have

$$\{F\}^* = \{Q\}^* = \{E\}^* = \{E\}. \tag{3.8}$$

Here, $\{*\}$ denotes the reachable set for trajectories and $\{*\}^*$ is the reachable set for
quasitrajectories.

Let $W = R^n \times T$ and $\Theta(k)$ be the hyperplane $t = k$. Also, denote $S(E, k) = \{E\} \cap \Theta(k)$ (the "time section of E"). One of the results shown by Zaremba [48] is
that $S(E, k)$ is a compact and connected set. Thus, the same property holds for the
sets of quasitrajectories of the fields $F$ and $Q$. A more detailed discussion on this
issue and other properties of the reachable sets can be found in Wazewski [45]. For
the field $E$, it is also known that if a point is accessible from the initial point $x(0)$ of
(3.5), then it is also accessible in optimal (minimal) time.

## 3.4 Differential Inclusions and Control Systems

The DIs are closely related to control systems. To see this, consider a dynamic system
(3.9)

$$\begin{cases} \dfrac{dx}{dt} = f(x(t), u(t), t), \ x(0) = x_0, \ u(t) \in C(x(t), t) \\ x \in R^n, , \ u \in R^m, \ C(x(t), t) \subset R^m, \end{cases} \tag{3.9}$$

where $x(t)$ is the n-dimensional system state, $u(t)$ is the m-dimensional control variable, and the set $C(x, t)$ represents the control restrictions. In the following, we will consider differential inclusions with the initial set reduced to a point $x_0$ in $R^n$.

Define a mapping $F$ as follows:

$$F(x, t) = \{z : z = f(x, u, t), u \in C(x, t)\}. \tag{3.10}$$

Using the set-valued function $F$ in the differential inclusion (3.5), we obtain the DI derived from the control system (3.9):

$$\frac{dx}{dt} \in F(x(t), t), \quad x(0) \in R^n. \tag{3.11}$$

The control system (3.9) and the DI (3.11) have the same quaitrajectories. In (3.11), the control variable does not appear explicitly. The function $f(x, *, t)$ is the following mapping:

$$f : C \to F, \text{ for each } x, t. \tag{3.12}$$

Define the bang-bang kernel of $C(x, t)$ as follows:

$$B(x, t) = \{u : u \in C(x, t), \ f(x, u, t) \in Q(x, t)\}. \tag{3.13}$$

The consequence of (3.8) is that we can use the bang-bang kernel $B$ instead of $C$ to obtain a control system with the same quasitrajectories. This means that we can hit or approximate any point inside the reachable set of (3.11), using the restricted control set $B$. The set $B$ contains less points than $C$. In many practical applications, $B$ can be reduced to a finite number of points. This permits us to use a simple bang-bang control instead of continuous control, with less complicated instrumentation.

As a consequence of the above remarks, we may conclude that given a differential inclusion, we can find the corresponding control system by parametrizing the function $F$ with a certain parameter $u$ (control variable). This is true in many cases. However, the parametrization problem is not so simple. Consult Aubin and Cellina [4], Chapter 1 Section 7, "Application: The parametrization problem." In that section, it is pointed out that the existence of continuous selection of $F$ is not sufficient to enable the parametrization of $F$. Fortunately, the mappings considered in the following are regular enough to permit parametrization.

### 3.4.1   Uncertainty Treatment

Models are frequently charged with some *uncertainty*. This may be caused by the errors in parameter estimation, or by the influence of the external, constant, or fluctuating control signals and disturbances. A common approach to uncertainty treatment is the use of random variables. This is also referred to as *stochastic models* that include such variables in the equations of model dynamics. However, observe that, to use random variables, we must know their probabilistic properties, like the expected value, variance, probability distribution, and others. These data may be difficult to obtain in practical applications. Instead, it is probable that we only know the limits where the variable can change. In this case, the statistical methods can hardly be applied or may provide wrong results. The point is that an *uncertain variable* need not be a *random variable*. The uncertainty may cause fluctuation, permanent constant changes, or unpredictanble tendencies. For example, this may be caused by natural disasters, or an erroneous information generated intentionally, like in the stock market models (remember also the law of Murphy).

Such uncertain factors are called *tychastic variables*, as defined in Aubin and Saint-Pierre [3]. The reachable sets may show the influence of such uncertain variables on the system behavior. Note that using tychastic variables, the *problem statement is deterministic*. We define the restrictions for tychastic variables and then calculate the shape of the reachable set of the model affected by them. This gives us insight into possible model uncertainty in the form of the RS. In the examples shown below, the external input signals can be treated as tychastic variables and the resulting RS images as the uncertainty regions.

## 3.5 Functional Sensitivity

The discussion about the functional sensitivity methods can be found in Chap. 4. Here, there are only short, preliminary remarks.

The classical, local sensitivity analysis (basic local version) uses the partial derivatives of the model output $Y$, with respect to components of an input vector (model parameters) $u = (u_1, u_2, \ldots, u_n)$, at a given point $u_0$:

$$\left| \frac{\partial Y}{\partial u_i} \right|_{u_0}. \tag{3.14}$$

The derivative is taken at some fixed point in the space of the input (hence, the "local" in the name of the analysis mode). The use of partial derivatives suggests that we consider small perturbations of the input vector, around the point of interest $u_0$. Consult Cacuci [8]. There are several kinds of sensitivity analysis. Consult, for example, scatter plots (Friendly and Denis [16]), regression analysis (Freedman [15] or Cook [10]), variance-based model, Sobol methods [37], the screening method (Morris [25]), or logarithmic gain, (Sriyudthsak et al. [38]).

The System Dynamics software offers tools for dynamic sensitivity analysis. Programs like Vensim or PowesSim include procedures that generate multiple model trajectories where the selected model parameters vary from one trajectory to another. However, in these packages, the parameters are constant along the trajectory. Our approach is different. As explained in the following sections, we treat the perturbations as functions of time. The main tool used here is differential inclusion.

In this book, the *local functional sensitivity* is defined as follows:

$$S_k = \left| \frac{\delta x_k}{\delta u} \right| \tag{3.15}$$

Note the difference between the conventional local sensitivity (3.14) and the functional sensitivity (3.15). The notation $\delta u$ denotes the variation of the function $u$, as defined in the calculus of variations (see Nearing [27] and Elsgolc [12]). A variational approach to sensitivity is also discussed in Arora [1], Mordukhovich [24], and Sriyudthsak [38]. A more detailed explanation is given in Chap. 4.

## 3.6  Differential Inclusion Solver

The original algorithm of the solver has been published in 2002 [32]. Here, we present the new version that includes the use of multiprocessing, improved accuracy, and graphical presentation of the results.

The first version of the differential inclusion solver (DI Solver) was presented two decades ago, see Raczynski [32]. We present this chapter because the publication of 2002 was only a conference paper and has been unnoted by many researches. The other reason is that during the last two decades important improvements to the algorithm and related software have been done. Also, what is new in the present article, is the application of the DI solver to a new concept of functional sensitivity analysis.

First of all, note that the main idea of the software is to scan the boundary, and not the interior of the reachable set (RS). Our point is that to explore the interior of the reachable set is an error. The reasons are as follows.

1. For example, in the case of three-dimensional state-time space, the area of the boundary of the reachable set grows with the square of the model size, while the volume of its interior increases as the cube of the dimension.

2. The properties of the RS boundary are perfectly known for many decades. This makes the RS calculation easier because the algorithm is based on already-proven and well-documented methods of the optimal control theory (see Lee and Markus [20], Pontryagin [30]).

The application of the optimal control methods for RS calculation has been re-invented several years after the original publication in 2002.

There are some other approaches to the RS determination, for example, Girard [17]. However, the method is limited to time-invariant linear systems. Matviychuk [22] proposes external ellipsoidal approximation method. We will not comment here on other numerous publications on this topic because nearly all of them propose less efficient methods and have been published after 2002 when the problem has been already solved. This remark does not refer to works that treat extensions of the problem and non-classic cases, like systems with time-delay or fuzzy logic elements. For more references related to reachable set calculations and an overview, consult Filippova [14].

The algorithm of the DI solver has been coded using the Embarcadero ® Delphi and requires that package to be installed on the user's machine. A limited, stand-alone ".exe" version of the solver is also available. Our main goal is RS determination and not optimization. The DI solver and the present problem statement should not be confused with the differential inclusions method used in the optimal control problems.

We are looking for the reachable set for a given initial condition. It is known that, with sufficient regularity assumptions, the reachable set is continuous with respect to the initial point or set, and it is a connected set for any fixed time instant. The reachable set need not be convex and may have a complicated shape. It might appear that a simple way to get the shape of the reachable set is to calculate a number of solutions to the following equation:

$$\frac{dx}{dt} = z(t), \tag{3.16}$$

for different $z(t)$, where the functions $z(t)$ are selectors of function $F$ of (3.5). It might appear that, by choosing $z(t)$ randomly, we can cover the inside of the reachable set with sufficient density and then estimate its shape. Unfortunately, this is not true, even if we select only $z(t)$ belonging to the boundary of $F$. We will call this method *simple or primitive shooting*, and compare it with the DI solver algorithm. A simple simulation shows that even in very simple cases the set of trajectories provided by primitive shooting (using any density function) is concentrated in some small region inside the reachable set and does not approach its boundary.

Using the DI solver, we explore the boundary of the reachable set, and not of its interior. With enough trajectories that belong to the boundary, the shape of RS can be estimated with reasonable accuracy. We should generate these trajectories in such a way that the density of points be nearly uniform on the boundary. This permits to avoid "holes" in the resulting graphical image.

To generate such "boundary trajectories" we use well-known methods of the control theory. If the field $F$ is not convex, but Lipschitzian, it is sufficient to estimate the reachable set for trajectories of the corresponding tendor field instead of the original field $F$. This can be easier because the tendor field contains few points, in many cases only isolated extremal points of the given set $F$. Recall that the fields $F$, $E$, and $Q$ (Fig. 3.2) have the same quasitrajectories and that for each quasitrajectory a regular trajectory exists nearby.

The Maximum Principle (Pontryagin [30]) states that the necessary condition for a trajectory to be optimal is to maximize the expression called Hamiltonian for each time instant over the time interval under consideration. In other words, the principle permits us to decompose the original optimization problem of maximization of a function into a set of problems of function maximization. The original optimization problem is as follows. Given a control system described by the Eq. (3.9), we look for an optimal control and the corresponding optimal trajectory that minimizes a given criterion (3.17) over the interval $I = [0, T]$.

$$J = \int_0^T f_0(t)dt. \tag{3.17}$$

If $f_0 \equiv 1$, then the trajectory is time-optimal, i.e., reaches the final point in optimal time. To define the Hamiltonian, we must define the conjugated vector $p \in R^n$ that satisfies (by definition) the following equations:

$$\frac{dp_i}{dt} = -\sum_{j=1}^n \frac{\partial f_j}{\partial x_i} p_j - \frac{\partial f_0}{\partial x_i}, i = 1, 2, \ldots, n, \tag{3.18}$$

where $f$ is the function of the right-hand side of (3.9). The necessary condition for a trajectory to be time-optimal is to entirely lay on the boundary of the reachable set. This means that we can suppose $f_0 \equiv 1$, and eliminate $f_0$ from the above equations.

The *Hamiltonian* function is defined as follows:

$$H = \sum_{j=1}^n p_j f_j. \tag{3.19}$$

The Maximum Principle states that the necessary condition for the trajectory to scan the RS boundary is that the control $u(t)$ maximizes the Hamiltonian over the interval $I$. This can be used to generate boundary trajectories of the differential inclusion. If the inclusion is given in the form of a control system (3.9), then we apply the principle directly. If it is given in the general form (3.5), we must parametrize the set $F$ and treat the parameter as the control.

The vector $p$ must satisfy the transversality conditions (Lee and Markus [20]). This provides the final condition for the conjugated vector. Consequently, to calculate an optimal trajectory with a given optimality criterion, the *two-point boundary value* problem must be solved. We know the initial condition for the state vector, but not for the vector $p$ that is defined at the final time.

In our case, we do not need to solve the two-point boundary condition problem. Observe that starting with *any* initial condition for the conjugated vector and maximizing the Hamiltonian on each time-step, a single forward integration of the Eqs. (3.9) and (3.18) provides a trajectory that lies on the boundary of the reachable set. Thus, we can choose the initial conjugate vector randomly, obtaining random final boundary points (and not the points inside the reachable set). We do not solve any particular optimization problem, but we are just looking for the trajectories that scan the boundary of the reachable set. The problem is how to generate the initial vector $p$ to cover the resulting final boundary set with a uniform density of points and to avoid holes in it.

The distribution $D$ used below is a probability distribution function defined inside the n-dimensional unit cube with a center at the origin of the coordinate system. The algorithm is as follows (the discrete-time version of the Maximum Principle is used).

The new, parallel version of the DI Solver algorithm is as follows.

*0. Define D as the uniform density function, set $x = x_0$.*

*1. Generate initial vector p according to the density D.*

*3. Launch several concurrent tasks on the available processors of the current machine, each of them, integrating the equations 3.9 and 3.18 over the interval I. In each trajectory, use the control u that maximizes the Hamiltonian at consecutive integration steps.*

*4. In each integration task, store the initial p and the whole trajectory in a consecutive record of a file.*

*5. Select the final point $x_k$ that lies in a region of the minimal density of points x, searching in the file where trajectories have been stored.*

*6. Modify the distribution D, increasing the probability density in a neighborhood of the point $p_k$ that corresponds to the point $x_k$.*

*7. If there are enough trajectories stored, then stop; otherwise, go to step 1.*

The difference between the original version of the solver and the present one is that the trajectory integration is now executed concurrently on multiple CPU processors, making the algorithm several times faster. By the term "region of minimal density of points" (point 5), we understand the spot of low-density of points on the image of the final reachable set or its two- or three-dimensional projection. This part of the algorithm is rather heuristic.

It is important to notice that, although we randomly generate the initial conditions for the conjugated vector, this algorithm has nothing to do with the simple shooting method, mentioned earlier or with simple random disturbances. We do not explore the interior of the RS but scan its boundary. In fact, the random generation of the initial vector $p$ can be replaced by other, deterministic, methods as well.

The stop condition is based on the user decision. As the result, we obtain a two-dimensional image that is a projection of the reachable set on a given plane. In the case of a second-order system, this image should be a closed curve; for systems of higher order, it will be a cloud of points in $R^n$. The user can stop the program if he/she recognizes the shape of the reachable set. A three-dimensional image of the reachable set can also be generated. Practical experiments show that this can be reached after integrating 500–1000 system trajectories. Anyway, this procedure may be difficult to use for models of higher order (more than 10, perhaps).

The maximization of the Hamiltonian (step 3) can be done using any maximization procedure. This procedure is not predefined because it may depend on each particular case. Note that here we reduce the original problem of solving a DI to some "sub-problems" that may not be easy to solve, but belong to well-known fields of optimization techniques. If the original system is linear with respect to the control vector and the restriction set is a multidimensional cube, then the maximization can be reduced to a simple scan over a finite number of points. We will not discuss here the methods of maximization of the Hamiltonian (step 3). There exists a huge literature on it in the field of control theory (Lee and Markus [20], Polak [29]).

An important question is if, and when, you really need the reachable set calculated by the DI solver. Obviously, if our system is of the first order, the determination of the reachable set is rather trivial and can be done easily without involving DIs. In some cases, when the modeled system is of higher order, but strongly damped, the extreme points of the RS can be calculated simply by applying the extreme values of control variables. However, in a general multidimensional case, the fixed extreme controls (extreme points of the control restriction set) do not correspond to the extreme or boundary points of the system reachable set. The RS becomes a complex multidimensional shape, not necessarily convex, with a boundary surface that may fold several times. Even in a simple two-dimensional case, the mapping from the permitted control set $C$ to the RS is highly irregular.

In some situations, the DI solver may fail. This occurs when the analyzed model includes stiff equations. This is normal, recall that most of the numerical methods for ordinary differential equations also fail for such type of equations, and the treatment of stiff systems requires special algorithms. The model stiffness is not always easily detected. The solver failures may, for example, occur when the model is of order four or higher, and includes parts that oscillate at high frequency. For example, the model of quarter-car suspension has the proper ("slow") oscillations of the spring-damper-mass part, and the stiff part that takes into account the tire elasticity and the mass of the wheel.

For models of order greater than two, the time section of the reachable set cannot be seen as a well-defined contour. For example, if the model is composed of a set of three equations, the final RS is a cloud of points distributed over a three-dimensional surface, like a balloon. If we look at a two-dimensional projection of this object, it might appear that some points lie inside the reachable set, which is not true. Rotating the image gives us a better understanding of the spatial distribution of points.

The algorithm was adopted for multi-processing. Note that the operations of integrating model trajectories can be executed concurrently. This accelerates the task due to the number of processors. Running on a quad CPU, we complete the task four times faster. Further improvements can be done using the GPU (Graphical Processing Unit) that may contain hundreds of processors. In this case, we could reach the velocity of RS calculation comparable with the speed of recent numerical methods for the ordinary differential equations. However, such implementations are hardware-dependent, and the practical applications are available mainly with the NVIDIA graphic cards.

As mentioned before, the DI solver runs over the Delphi package. A new, standalone exe version of the solver has been developed, that does not require Delphi. This version has its own compiler for mathematical expressions. However, this compiler is rather slow compared that of Delphi and does not use multiprocessing. This version of the solver can be used for simple examples and not for complicated models.

### 3.6.1  Example: A Second-Order Model

Consider a simple non-linear model of the second order:

$$\begin{cases} \dfrac{dx_1}{dt} = x_2 \\[2mm] \dfrac{dx_2}{dt} = 1 - 0.2u_1 - x_1 - 0.1u_2(x_2 + 2.3x_2^2), \end{cases} \tag{3.20}$$

where $u_1$ and $u_2$ are uncertain parameters. Let the parameter $u_1$ fluctuate between $-0.2$ and $+0.2$, and parameter $u_2$ fluctuate between 0.025 and 0.175. The initial conditions are $x_1 = x_2 = 0$, and the final simulation time is equal to 10. Figure 3.3 shows the 3D image of the model RS. The three axes of the plot represent $x_1$, $x_2$ and the time. The image was generated by our DI solver. It can be seen that even for relatively small perturbations $u_1$ and $u_2$, the deviation of the state vector may be quite big.

Figure 3.4 part A depicts the comparison of the functional sensitivity to the conventional sensitivity analysis provided in some system dynamics packages. The contour indicates the boundary of the reachable set for the model (3.20) at $t = 10$. These are end points of about 2000 boundary-scanning trajectories. A small black region marked with X is the result of the "Vensim-like" sensitivity analysis, where the parameters are constant along each trajectory. The region X was obtained by

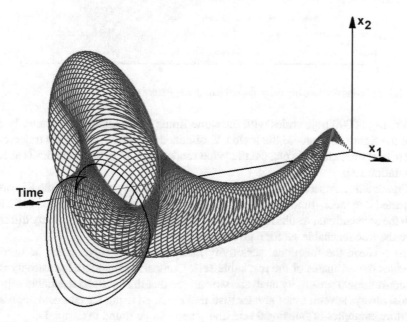

**Fig. 3.3** Reachable set of model (3.20). The image of the reachable set produced by the DI solver

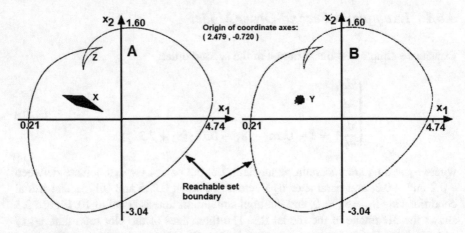

**Fig. 3.4**  Final contour of the reachable set. Comparison with the simple shooting

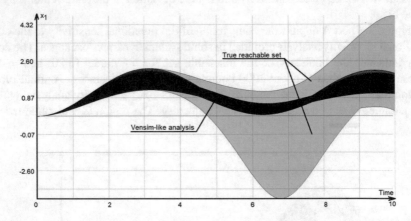

**Fig. 3.5**  Functional sensitivity set projected into the $x_1-time$ plane

generating 50000 trajectories with the same limits of uncertain parameters. In part B of the figure, we can see the region Y, obtained by generating 50000 trajectories where the parameters can change the value randomly, within the same limits at each integration step.

Figure 3.5 shows a side view of the same reachable set produced by the DI solver, projected into the $x_1$-time plane. Note that the functional sensitivity region coincides with the conventional results for the initial time interval [0, 35], but it is very different from the true reachable set for a greater time interval.

To perform the functional sensitivity analysis, the DI solver must be used. It provides the estimates of the reachable sets. Comparing functional sensitivity with the conventional sensitivity analysis, we can see that the obtained reachable sets are almost always several times greater than that obtained using the classical approach.

More examples of functional sensitivity sets can be found in Chap. 4.

## 3.7 Discrete Differential Inclusions

In the discrete case of differential inclusion (or rather *difference inclusion*), we take an important assumption: The steps (increments) of time discretization, as well as the model state, are finite. **No operations or considerations that involve passing with these increments to zero and looking for some limit values or properties are considered**.

In model building, we can frequently find the concepts of "discrete version" and "continuous version." For example, a system dynamics model may be described by differential equations, as well as by difference equations. In the latter case, a finite time-step is used, and the model trajectory is calculated by a simple numerical algorithm similar to Euler's method. In many cases, it is supposed that, by passing with the time-step to zero, we obtain the continuous model of the same real system. As stated before, we do not permit such operation in this section. In other words, we treat the discrete (time and state) models as something different from the continuous models, and no "passing to time-step zero" operations are possible. This does not mean that there cannot exist some analogies between our discrete and continuous models. We just treat them as different things.

There are some publications on discrete differential inclusions (DDIs), see for example, Veliov [42], who discusses the approximations of differential inclusions by discrete-time inclusion. However, the topic of the present paper is quite different. We do not treat with approximations of the continuous version. In our approach, the inclusion is discrete both in the time domain and in the state space. This is a "strongly" discrete approach.

The DDIs can be treated as some kind of cellular automata. Recall that a cellular automata consists of a collection of cells on a grid in space. Each cell can evolve through consecutive time-steps, due to a given rule of change. This rule may be constant or variable in time and normally depends on the state of the cell and its neighborhood. However, the DDI approach is not exactly a cellular automata. In the cellular automata, each cell has its state. While advancing in time, the cell may change its state, for example, from "0" to "1." In a DDI, the cells represent possible states of the model, and the model trajectory jumps from one cell to another, changing the state of the model and not of the cells.

The cellular automata evolve, creating interesting patterns and sometimes quite unexpected images. This pattern creation and moving clusters resemble biological systems and sometimes are treated as "artificial life." The cellular automata are not the subject of this book, so we will not discuss them here. Consult, for example, von Neuman [41] or Wolfram [47].

In a DDI we have the following elements: the *space of states, two- or multi-diensional grid of states, the initial set the model movement starts from, the rule of change and additional state restrictions*. Compared to continuous DIs, the problem statement is similar. The rules of change of a DDI correspond to the right-hand side of the DI (in the continuous case it is the set where the derivative of motion must belong). The DDIs may be a useful tool in system simulation and decision-making. Any trajectory of a DDI is a sequence of *decisions* taken about the system movement. If we associate each state with an object function (e.g., a cost function), then we may consider an optimization problem for a decision-making system.

Consider a DDI in which states are vertices of a two-dimensional grid of states. The initial condition is given by one vertex of the grid. This restriction to two-dimensional state space is not relevant. The problem statement can be n-dimensional, as well. We limit the examples to two-dimensional cases because this permits us to generate images that are clear and easy to interpret. Note that this grid is fixed, and the spacing cannot be changed and it cannot approach a continuum. This problem is conceptually completely different from the continuous version. We have no continuous or differentiable functions and the concept of a derivative does not exist here. The optimization problem cannot be treated by Pontryagin's maximum principle, even if this is applied in a discrete-time version, consult Pontryagin [30]. In our case, the model state is also discretized and belongs to a space of isolated points. Recall the very fundamental difference between a continuous and discrete state space. Consider the space of real numbers. In this space, we can define a closed or open set. Now, consider the set of all integer numbers. Recall that this set has no interior, in other words, it has different topological properties.

Here, we describe an algorithm that calculates the reachable sets of DDIs and provides optimal trajectories. Note that in the process description we don't use much mathematics. The optimality of the trajectories generated in this process needs no mathematical proof.

### 3.7.1  Reachable Set, Optimal Trajectory

Let $x_k = (x_{k1}, x_{k2})$ be the actual model state, where $k$ represents the time. Both state and time are discrete, so we suppose that $k = 0, 1, 2, \ldots$. The possible values of each state component are integer numbers. While advancing in time, the new state is calculated as equal to

$$x_{k+1} = x_k + dx_k, \quad dx_k = (dx_{k1}.dx_{k2}) = f(x_k., u_k, k). \qquad (3.21)$$

Model specification defines the increment function $f(x_k, u_k, k)$, where $u_k$ is an external control vector that influences the state changes. In this discrete model, the vector $dx$ can be interpreted as a (discrete) decision that tells "where to go." Vector $u$ can also be interpreted as the uncertainty in the decision-making process. Let the control $u$ be restricted so that

$$u_k \in C(x_k, k). \qquad (3.22)$$

For $k = 0, 1, 2, \ldots, K$, with given initial state $x_0 = (x_{01}, x_{02})$ and permissible set $C$. $K$ is a given final model time. The formula (3.21) and (3.22) define the Discrete Decision Inclusion (DDI). Note that, similar to the continuous DI, the right-hand side of the first equation of (3.21) is a set. However, the model state $x$, as well as the increments $dx$ take discrete values.

A trajectory of the DDI (3.21) is a succession of states that satisfy Eq. (3.21) for $k = 0, 1, 2, 3, \ldots, K$.

The reachable (or attainable) set (RS) of the DDI (3.21) is the union of graphs of all possible model trajectories with state $x_0$ belonging to the initial set, and final time $K$. In the following, the initial condition will be a one-point set, denoted as $x_0$. In other words, this is the set of all possible points $x_k$, where $x_k$ belongs to one of the model trajectories.

By an optimal trajectory we mean a trajectory that satisfies some optimality criterion at the final time instant $K$. It may be, for example, the task to maximize one of the components of the final state at $time = K$. Of course, optimality can be defined in many ways, as is done in the optimal control theory. Here, we limit us to the requirement that the final model state belongs to the boundary of the reachable set at $time = K$. The optimality problem is not the main point of this chapter. There are many works on this topic, see for example, Blackwell [7], where we can find the description of the discrete version of dynamic programming method of Bellman [5]. See also Elmaghraby [11] and Ibaraki [19]. Here, we focus on the determination of reachable sets, and the simple way to obtain optimal trajectories is a "side" product.

We should define the notion of the set boundary. Remember that our reachable set is a cloud of discrete points of the two-dimensional grid (G) with spacing equal to one. So, from a topological point of view, this set has no interior and no boundary. However, for our needs, we will consider a state point as a boundary point, if it has less than four neighbors (at distance one) in the grid G; otherwise, it is an interior point.

Now, consider the problem of reaching a particular final state, starting with a given initial point. Suppose, for example, that we have $K = 30$ time-steps, the component $u_1$ has five possible values, and $u_2$ has two. So, at each time-step, we have ten possible control vectors. This means that we have $10^{30}$ combinations of control variables for each trajectory. Obviously, it is hardly possible to generate this number of trajectories in reasonable computing time, and select the optimal one. As for the shape of the reachable set, one might suppose that, by generating randomly a sufficiently large number of model trajectories, we can scan the interior of the reachable set and assess its shape. Unfortunately, this is not the case. Such trajectories form a small cluster inside the RS, and the probability that they approach the boundary is practically null.

So, instead of generating trajectories, we generate consecutive "slices" of the RS. By a slice, we understand the set of points of the reachable set for a given time instant.

Let the number of possible values of $u$ be fixed for each time-step and equal to $N$. We start at $K = 0$, at the initial point $x_0$, and generate the set of states for $K = 1$. this will be a set of $N$ points. Denote this set (slice) by $S_0$. Now, repeat this starting from the points of $S_0$. We obtain a new slice $S_1$ that can include up to $N^2$ points. Note that the states are discrete so that some of the new points in $S_1$ may be located in the same grid vertex. If we eliminate these double points, we have a slice with less than $N^2$ points. This point reduction accelerates the process significantly. For example, suppose that the limits both $u_1$ and $u_2$ change between 0 and 3, $dx_1 = u_1, dx_2 = u_2$. Advancing in time, we get the consecutive slices with 1, 16, 49, 100, 169, 256.... points. Now, repeating the process without eliminating multiple points, the number of points in consecutive steps is 1, 16, 256, 4096, and 65526 for only the first four time-steps. This point reduction is an advantage of the fact that we have discrete, and not continuous model states. It can be seen that while eliminating repeated points, the process is computationally tractable and the reachable set can be calculated in a reasonable time. In such types of models, the number of time-steps is not very big. If we require, for example, 2000 time-steps, this would suggest that we intend to pass with the size of the time-step to zero (approximate a continuous process), which is not permitted in our problem statement. In the examples shown below, the number of time-steps does not exceed 50.

Any trajectory that terminates in a given endpoint (including the optimal one) can be easily retrieved. In each time-step, for each new point we store the actual coordinates, coordinates of the previous point it comes from, and the corresponding control $u$. These data permit to restore the trajectory, going backward in time. Note that this is an additional, "postmortem" feature.

### 3.7.2 Example 1

Consider a model described by Eq. (3.21), where the decisions are as follows.

$$\begin{cases} dx_1 = x_2 + u_1 \\ dx_2 = 5 - r(0.3x_1 - 0.45x_1) + 2u_2. \end{cases} \tag{3.23}$$

Here, $(x_1, x_2)$ is the actual state, $(u_1, u_2)$ is the external control or decision uncertainty and $r(*)$ is the "round" function that returns the integer value nearest to the argument. Remember that $dx_1$ and $dx_2$ are integer numbers. The initial state of x is equal to $(0, 0)$. The external control is limited as follows

$$u_1 \in \{-1, 0, 1.2\}, \quad u_2 \in \{-1, 0\} \tag{3.24}$$

Figure 3.6 shows a 3D image of the reachable set for $K = 20$ time-steps, shown from two different view angles. In the figure, each small sphere represents system state. Each gray sphere represents a state point that has four near neighbors in the grid, in the consecutive slice. White balls are the boundary points, i.e., the points that have less than four such neighbors. The irregularity of the RS boundary is the consequence of the time and state discretization. The calculation of this reachable set takes less than two seconds on a PC, including graphics generation.

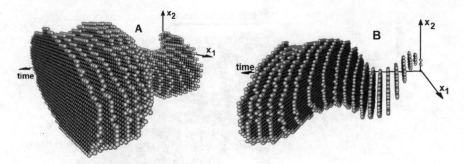

**Fig. 3.6** Reachable set for example 1, shown from different view angles

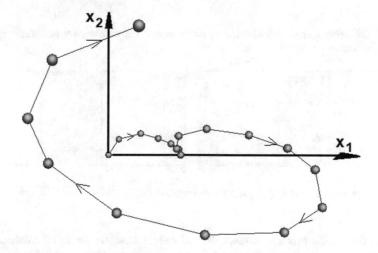

**Fig. 3.7** Optimal trajectory for example 1, viewed from the time-axis direction

As mentioned before, an additional feature of the program is optimization. After generating the reachable set, the user can select any point from the final slice. The trajectory that reaches this point and the corresponding control are displayed.

In Fig. 3.7, we can see the optimal trajectory that maximizes the final value of $x_2$. The 3D image of the reachable set with the optimal trajectory is shown in Fig. 3.8. Note that the graph of the whole trajectory belongs to the boundary of the reachable set. This property is well known in the optimal control theory, for the continuous case, but here it appears in a very natural way.

The optimality property of the trajectories obtained by the algorithm described here needs no mathematical proof. Simply, the reachable set we generate contains all possible system states. Then, the user selects the required point (extremal point or any other), and the program shows the corresponding trajectory and control. Figure 3.9 shows the two controls that correspond to the optimal trajectory. Obviously, to reach the maximal value of $x_2$, we cannot maintain the control $u_2$ at its maximal value in all time-steps.

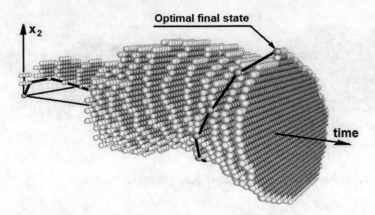

**Fig. 3.8** Reachable set and optimal trajectory for example1. The problem is to maximize the value of $x_2$ at the final model time

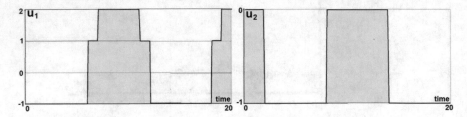

**Fig. 3.9** Optimal controls: $u_1$ (left part) and $u_2$ (right)

Note that in the process of reachable set calculation, we never calculate model trajectories to assess the shape of the RS (we can retrieve them from the stored data, and not integrate in time). Instead, the algorithm finds the consecutive time sections (slices) of the RS.

### 3.7.3  *Example 2*

Consider the following model:

$$\begin{cases} dx_1 = r(0.2x_1 - 0.015x_1x_2) + u_1 \\ dx_2 = r(-0.3x_2 + 0.01x_1x_2) + u_1, \end{cases} \tag{3.25}$$

where $-1 \leq u_1 \leq 0,\ 0 \leq u_2 \leq 1$ Like in Eq. (3.7), $(x_1, x_2)$ is the actual state, $(u_1, u_2)$ is the external control or decision uncertainty and $r(*)$ is the "round" function that returns the integer value nearest to the argument. The initial conditions are as follows: $x_1(0) = 20$, $x_2(0) = 5$, final time $K = 25$.

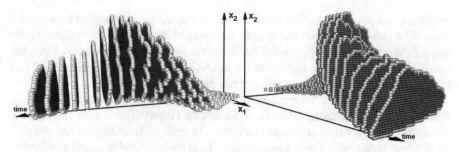

**Fig. 3.10**   A 3D image of the RS for model (3.25)

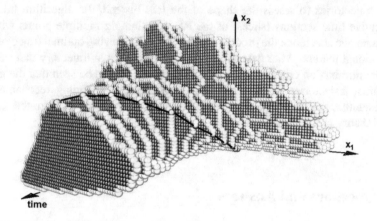

**Fig. 3.11**   The RS and the optimal trajectory for model (3.25)

The image of the RS for this model is shown in Fig. 3.10, from two different view angles. Note that in this case, the slices of the RS can shrink while advancing in time (Fig. 3.11).

## 3.8   Conclusion

In this chapter we discuss models based on differential inclusions, instead on the ODEs. Differential inclusions are not very popular among the simulationists, though they are known for more than 90 years. More applications can be found in works on optimization methods. However, the images of the reachable sets of the DIs can hardly be found in the literature. This chapter intents to show a new type of model. Another topic is functional sensitivity that is a much dynamic approach, compared to conventional sensitivity analysis.

New DI solver contains considerable improvements compared to the original version. Running on multiple processors, it can calculate the shape of the reachable set in several seconds. A possible application of the GPU hardware may accelerate the solution hundreds of times. The graphical result display was redesigned to provide a more clear and easier to interpret images. An application to the functional sensitivity analysis shows that, in fact, the RS for dynamic systems is, at the same time, the sensitivity set. This can be applied in the analysis and design of robust control and other problems, subject to uncertain disturbances.

Discrete version of differential inclusion was presented. This may be useful in multi-step decision-making process. It should be emphasized that in the discrete inclusions, we treat time and state discretization with finite time-step and the grid of discrete states with finite spacing. No "pass to zero" with the discretization steps is considered. In the process of reachable set calculation, we never calculate model trajectories to assess the shape of the RS. Instead, the algorithm finds the consecutive time sections (slices) of the RS. Eliminating multiple points with the same state, we accelerate the process considerably. Retrieving optimal trajectories is an additional feature. After determining the reachable set, a trajectory that satisfies a given optimization criterion can be easily retrieved. It can be seen that the graphs of optimal trajectories lie at the boundary of the corresponding reachable sets. This resembles the similar property of optimal trajectories known from the optimal control theory in a continuous case.

## 3.9   Questions and Answers

**Question 3.1** What is a *differential inclusion*?

**Question 3.2** What is a *selector* of a field $F(x, t)$?

**Question 3.3** What is a *trajectory* of a differential inclusion?

**Question 3.4** What is theit reachabke or it attainable set?

**Question 3.5** What is the difference between it uncertain and *random* parameter or variable?

**Question 3.6** How the *differential inclusion solver* works?

**Question 3.7** What is the difference between the *functional sensitivity* of models and the conventional sensitivity definition?

**Answers**

**Answer 3.1** A differential inclusion is given in the form of the condition $dx/dt \in F(x, t)$, where $F(x, t)$ is a set, $x$ is the state vector and $t$ is the time. It is also called *differential equation with multi-valued right-hand side*.

**Answer 3.2** A selector of the field $F(x, t)$ is a function $z(t) \in F(x, t)$ over the time interval of interest.

**Answer 3.3** A *trajectory* of a differential inclusion is an absolutely continuous function that satisfies the condition $dx/dt \in F(x, t)$ over the time interval of interest.

**Answer 3.4** The it reachabke or it attainable set is the union of the graphs of all trajectories of a differential inclusion with a given initial set in the state space.

**Answer 3.5** A random variable has certain probabilistic properties like the average, variance, and others. The *uncertain* variable need not have such characteristics; it has just a uncertain value. Such variables are also known as *tychastic* variables.

**Answer 3.6** The *differential inclusion solver* generates trajectories that scan the *boundary*, and not the interior of the reachable set.

**Answer 3.7** The *functional sensitivity* provides the reachable sets or the model trajectory related to parameters or disturbances that fluctuate in time. It is defined in terms of the *variational calculus*.

# References

1. Arora JS, Cardoso JB (2012) Variational principle for shape design sensitivity. Aerosp Res Cent 30(2):538–547. https://doi.org/10.2514/3.10949
2. Aubin JP, Chen L, Dordan O (2014) Tychastic measure of viability risk. Springer International Publishing. 978-3-319-08128-1
3. Aubin JP, Saint-Pierre P (2005) A tychastic approach to guaranteed pricing and management of portfolios under transaction constraints. Centro stefano franscini, ascona. In: Seminar on Stochastic Analysis, Random Fields and Applications V. https://doi.org/10.1007/978-3-7643-8458-6_22
4. Aubin JP, Cellina A (1984) Differential inclusions. Springer Verlag, Berlin. ISSN 978-3-642-69514-8, https://doi.org/10.1007/978-3-642-69512-4
5. Bellman R (1957) Dynamic programming. Princeton University Press, Princeton, NJ
6. Berkovitz LD (1964) Variational approach to differential games. In: Advances in game theory. Princeton Univ. Press, Princeton NJ
7. Blackwell D (1962). Discrete dynamic programming. https://doi.org/10.1214/aoms/1177704593
8. Cacuchi DG (2003) Sensitivity and uncertainty analysis: theory. I. Chapman & Hall, ISBN/ISSN 1584881151. https://doi.org/10.1201/9780203498798.
9. Chen L, Wang X-, Min Y, Li G, Wang L, Qi J (2020) Modelling and investigating the impact of asynchronous inertia of induction motor on power system frequency response. Int J Electr Power Energy Syst 117. https://doi.org/10.1016/j.ijepes.2019.105708
10. Cook RD, Weisberg S (1982) Criticism and influence analysis in regression. Sociol Methodol 13:313–461
11. Elmaghraby S (1970) The concept of state in discrete dynamic programming. J Math Anal Appl 29(3):523–557. https://doi.org/10.1016/0022-247X(70)90066-1
12. Elsgolc LD (2007) Calculus of variations. Dover Books on Mathematics. ISBN/ISSN 978-0486457994
13. Filippov AF (1967) Classical solutions of differential equations with multivalued right hand. SIAM J Control 5:609–621
14. Filippova TF (2017) Estimates of reachable sets for systems with impulsive control, uncertainty and nonlinearity. The Bulletin of Irkutsk State University. Ser Math 19:205–216

15. Freedman DA (2005) Statistical models: theory and practice. Cambridge University Press
16. Friendly M, Dennis D (2005) The early origins and development of the scatterplot. J Hist Behav Sci 41(2):103–130. https://doi.org/10.1002/jhbs.20078
17. Girard A (2005). Reachability of uncertain linear systems using zonotopes. https://doi.org/10.1007/978-3-540-31954-2-19
18. Halmos PR (1960) Naive set theory. The university series in undergraduate mathematics. van Nostrand Company, Princeton, NJ
19. Ibaraki T (1973) Solvable classes of discrete dynamic programming. J Math Anal Appl 42(3):642–693. https://doi.org/10.1016/0022-247X(73)90283-7
20. Lee EB, Markus L (1967) Foundations of optimal control theory. Wiley, New York. ISSN 978-0898748079
21. Marchaud A (1934) Sur les champs de demi-cones et les equations differielles du premier ordre. Bulletin de la Societe mathematique de France, 62, Societe mathematique de France
22. Matviychuk C (2017) Ellipsoidal estimates of reachable sets of impulsive control problems under uncertainty. https://doi.org/10.1063/1.5007411
23. Mordukhovich, (2005) Sensitivity analysis for generalized variational and hemivariational inequalities. Adv Anal 305–314. https://doi.org/10.1142/9789812701732_0026
24. Mordukhovich BS (1997) Optimal control of nonconvex differential inclusions. Report IIASA. Raport, Institute for Applied Systems Analysis, Vienna, http://pure.iiasa.ac.at/5261
25. Morris MD (1991) Factorial sampling plans for preliminary computational experiments. Technomet 33:161–174. https://doi.org/10.2307/1269043
26. Mujal-Rosas R, Orrit-Prat J (2011) General analysis of the three-phase asynchronous motor with spiral sheet rotor: Operation, parameters, and characteristic values. IEEE Trans Ind Electron 58(5):1799–1811. https://doi.org/10.1109/TIE.2010.2051397
27. Nearing J (2010) Mathematical tools for physics. In: Petrosjan L, Zenkiewicz NA (eds) Game theory. A book. World Scientific Publishing Co., Inc
28. Plis A (1961) Remark on measurable set-valued functions. Bulletin de Academie Polonaise des Science—Serie des Sciences Mathematiques. Astron Phys 9(12):857–859, Warszawa
29. Polak E (1971) Computational methods in optimization. Academic Press, New York. /ISSN 0125593503
30. Pontryagin LS (1962) The mathematical theory of optimal processes. Wiley Interscience, New York
31. Raczynski S (2011) Uncertainty, dualism and inverse reachable sets. Int J Simul Model 10(1):38–45, ISBN/ISSN ISSN 1726-4529
32. Raczynski S (2002) Differential inclusion solver. In: Conference paper: International conference on grand challenges for modeling and simulation, the society for modeling and simulation int., San Antonio TX
33. Raczynski S (1986) Some remarks on nonconvex optimal control. J Math Anal Appl 118(1):24–37. https://doi.org/10.1016/0022-247X(86)90287-8
34. Raczynski S (1984) On some generalization of "Bang-Bang" control. J Math Anal Appl 98(1):282–295. https://doi.org/10.1016/0022-247X(84)90295-6
35. Saunders Mac Lane (1998) Categories for the working mathematician. Springer (Graduate Texts in Mathematics), ISBN/ISSN 0-387-98403-8
36. Sentis R (1978) Equations diferentielles a second membre mesurable. Bollettino dell Unione Matemat Ita 15(B):724–742, ISBN/ISSN 1972-6724
37. Sobol I (1993) Sensitivity analysis for non-linear mathematical models. Math Model Comput Exp 1:407–414
38. Sriyudthsak K, Uno H, Gunawan R, Shiraishi F (2015) Using dynamic sensitivities to characterize metabolic reaction systems. Math Biosci 269:153–163
39. Turowicz A (1963) Sur les zones d'emision des trajectoires et des quasitrajectoires des systemes de commande nonlineaires. Bulletin de l'Academie Polonaise des Science—Serie des Sciences Mathematiques. Astron Phys 11(2), Warszawa
40. Turowicz A (1962) Sur les trajectoires et quasitrajectoires des systemes de commande nonlineaires. Bulletin de lAcademie Polonaise des Science—Serie des Sciences Mathematiques. Astron Phys 10(10), Warszawa

41. vonNeuman J. 1951) The general and logical theory of automata. In: Cerebral mechanisms in behavior—the hixon symposium, Wiley, New York
42. Veliov VM (1989) Approximations of differential inclusions by discrete inclusions, Working paper, International Institute for Applied Syetem Analysis, a-2361, Laxemburg, Austria, https://core.ac.uk/reader/33894798
43. Wazewski T (1963) On an optimal control problem differential equations and their applications. In: Proceedings of the conference held in Prague, publishing house of the Czechoslovak academy of sciences, Prague
44. Wazewski T (1962) Sur une genralisation de la notion des solutions dúne equation au contingent. Bulletin de lÁcademie Polonaise des Science—Serie des Sciences Mathematiques. Astron Phys 10(1)
45. Wazewski T (1962) Sur les systemes de commande non lineaires dont le contredomaine de commande nést pas forcement convexe. Bulletin de lÁcademie Polonaise des Science—Serie des Sciences Mathematiques. Astron Phys 10(1)
46. Wazewski T (1961) Sur une condition equivalente a léquation au contingent. Bulletin de lÁcademie Polonaise des Science—Serie des Sciences Mathematiques. Astron Phys 9(12)
47. Wolfram S (1984) Universality and complexity in cellular automata. Phys D: Nonlinear Phenomena 10(12):1–35. Elsevier, https://doi.org/10.1016/0167-2789(84)90245-8.
48. Zaremba SK (1936) Sur les equations au paratingent. Bull Sci Math 60

# Chapter 4
# Functional Sensitivity Applications

## 4.1 Introduction

The concept of *functional sensitivity* was mentioned in Chap. 3. Now, let us discuss this kind of analysis with more detail and examples. Unlike the classic sensitivity definition, we define the functional sensitivity in terms of variational calculus. It is pointed out that the non-local functional sensitivity is given in the form of the system reachable set. The solution to the functional sensitivity is shown, using the differential inclusion solver that calculates and displays the system reachable set. Comparison between functional sensitivity and the classical approach is done.

The topic of this chapter is the concept of functional sensitivity and reachable sets, and not the particular models we use.

First, let us express some remarks on the conventional sensitivity concepts (SA). There are many kinds of such analysis, as listed below.

**Scatter plots** represent a useful tool in the SA. Plots of the output variable against individual input variables are displayed. This gives us a graphical view of the model sensitivity. View Friendly and Denis [13] for more detail.

**Regression analysis** is a powerful tool for sensitivity problems. It allows us to examine the relationship between two or more variables of interest. The method is used to model the relationship between a response variable and one or more external variables or perturbations. Consult, for example, Freedman [12] or Cook [8].

**Sobol method** and **Screening** are useful tools in the variance-based modeling. It decomposes the variance of the output of the model or system into fractions which can be attributed to the input or sets of inputs. Thus, we can see which variable is contributing significantly to the output uncertainty in high-dimensionality models. For more detail, see Morris [18] and Sobol [24].

**Logarithmic gain** is a normalized sensitivity defined by the percentage response of a dependent variable to an infinitesimal change in an independent variable. In dynamical systems, the logarithmic gain can vary with time, and this time-varying sensitivity is called dynamic logarithmic gain. This concept is used in dynamic sen-

© The Author(s), under exclusive license to Springer Nature Switzerland AG 2022     107
S. Raczynski, *Models for Research and Understanding*, Simulation Foundations,
Methods and Applications, https://doi.org/10.1007/978-3-031-11926-2_4

sitivity analysis, where the core model is a dynamic system, described by ordinary differential equations. Consult Sriyudthsak et al. [25].

The **System Dynamics** software (Forrester [11]) offers tools for dynamic sensitivity analysis. Programs like Vensim or PowesSim include procedures that generate multiple model trajectories when the selected model parameters vary from one trajectory to another. However, in these packages, the parameters are constant along the trajectory.

Our approach is different. As explained in the following sections, we treat both the perturbations and uncertain model parameters as functions of time. Instead of the classical sensitivity concept, we use the functional sensitivity, defined in the following sections. The main tool used here is the differential inclusion solver that calculates model reachable sets (see Raczynski [23]).

## 4.2  Functional Sensitivity

### 4.2.1  Differential Inclusions

Differential inclusions (DI) has been defined and discussed in Chap. 3. For reader convenience, let us repeat here some basic concepts. Consider a dynamic model given in the form of an ordinary differential equation (state equation)

$$\frac{dx}{dt} = f(x, u, t), \tag{4.1}$$

where $x \in R_n$ is the state vector, $t$ is the time, $f$ is a vector-valued function and $u \in R_m$ is an external variable, called *control variable* in the automatic control theory. Suppose that the values of the control $u$ are restricted so that $u(t) \in C(t)$, where $C(t) \subset R_m$ is a subset of the $R^m$ space. For each fixed $x$ and $t$, the function $f$ maps the set $C$ (all possible values of $u$) into a set $F \subset R^n$. In this way, we obtain the following condition, that defines the corresponding differential inclusion.

$$\frac{dx}{dt} \in F(x, t) \tag{4.2}$$

A function $x(t)$ that satisfies (4.2) is a trajectory of the differential inclusion. The union of the graphs of all trajectories, over a given time interval and given initial condition, is called the *reachable set* of (4.2).

### 4.2.2  Sensitivity Analysis

The classical, local sensitivity analysis (basic local version) uses the partial derivatives of the model output $Y$, with respect to components of an input vector (model parameters) $u = (u_1, u_2, ..., u_n)$, at a given point $u_0$ :

$$\left.\left|\frac{\partial Y}{\partial u_i}\right|\right._{u_0}. \tag{4.3}$$

The derivative is taken at some fixed point in the space of the input (hence the "local" in the name of the analysis mode). The use of partial derivatives suggests that we consider small perturbations of the input vector, around the point of interest $u_0$. Consult Cacuci [6].

Consider a dynamic model described by an ordinary differential equation

$$\frac{dx}{dt} = f(x, u, t), \tag{4.4}$$

where $x = (x_1, x_2, ..., x_n)$ is the state vector, $u = (u_1, u_2, ..., u_m)$ is the perturbation (parameters, control) vector, and $t$ is the time. We have $x \in X, u \in U, f : X \times U \times R \rightarrow X$. Here, $X$ is the state space, $U$ is the control space, and $R$ is the real number space. We restrict the considerations to the case $X = R^n, U = R^m, R = R^1, R^k$ being the real Euclidean k-dimensional space. Let $t \in I = [0, T]$, and $G$ be the space of all measurable functions $u : I \rightarrow R^m$.

Now, consider a variation $\delta u$ of $u$ and a perturbed control $u^*$. The variation is a function of time, so that $u'(t) = u(t) + \delta u(t) \ \forall \ t \in I$ (prime mark is not the time differentiation). The solution to (4.4) over $I$, with given initial condition $x = x_0$ and given function $u(*)$, will be called a trajectory of (4.4). Thus, any component $x_k$ of the final value of $x(t)$ depends on the shape of the whole function $u(*)$. In other words, $x_k(t) = x_k(t)[u']$ is a functional (not a function) of $u'(*)$. Unlike a function, in our case, the functional is a mapping from the space $G$ to $R$. Denote $\delta x_k = x_k[u + \delta u] - x_k[u] = x_k[u'] - x_k[u]$.

In this book, the *local functional sensitivity* is defined as follows:

$$S_k = \left|\frac{\delta x_k}{\delta u_0}\right|. \tag{4.5}$$

Note the difference between the conventional local sensitivity (4.3) and the functional sensitivity (4.5). The notation $\delta u$ denotes the variation of the function $u$, as defined in the calculus of variations (see Nearing [20] and Elsgolc [9]). A variational approach to sensitivity is also discussed in Arora [2], Mordukhovich [17], and Sriyudthsak [25].

The term (4.5) defines a local property of the trajectory $x_k(t)$. Here, we are interested rather in the response of the model to perturbations that are not necessarily small. We will not enter in the methodology of the variational calculus. Our task is to define the functional sensitivity as the set of the graphs of all trajectories of (4.4), where $u = u_0 + \Delta u$. Here, $\Delta u(t)$ is a limited perturbation, not necessarily small. Considering the control system (4.5), this is equivalent to say that $u(t)$ belongs to the set of restrictions $C(x, t), u(t) \in C(x, t), \ \forall \ t \in I$. Here, $C(x, t)$ is a subset of $R^m$. When $u$ scans all possible values inside the set $C$, then the right-hand side of (4.4) defines a set-valued function. This way, (4.5) with disturbed control also defines the corresponding differential inclusion.

The functional sensitivity defined this way is non-local. We do not use the term "global," because this is not a global property of the model. We just do not require the perturbation to be small.

Models with uncertain parameters and control systems are closely related to differential inclusions. Consider a model defined as follows (Eq. 4.6):

$$\frac{dx}{dt} = f(t, x(t), u(t)), \quad x(t) \in R^n, \quad u(t) \in C(x, t), \quad t \in I = [t_0, T], \quad t_0 < T,$$

(4.6)

where $R^n$ is the real n-dimensional Euclidean space, $x \in R^n$, $u \in R^m$, $t$ is a real variable representing the time, and $C$ represents the restrictions for variable $u$. Equation (4.6) may represent a control system with control variable $u$, as well as a model with uncertain variable parameters $u$.

When $u$ scans all possible values in $C$, then the right-hand side of (4.6) scans the values inside a set $F$. This defines the corresponding differential inclusion, as follows:

$$\frac{dx}{dt} \in F(t, x).$$

(4.7)

Here, $F(t, x) = \{z : z = f(t, x, u), \ u \in C(x, t)\}$. This way, we obtain a differential inclusion (4.7). More detailed assumptions and a comprehensive survey on differential inclusions can be found in Aubin and Cellina [1].


## 4.3  Differential Inclusion Solver

The description of the differential inclusion solver is included in Chap. 3 Here, we only recall the main features. The basic version of the DI solver is not new. It was published in Raczynski [23].

In few words, the Di Solver generates a series of DI trajectories that scan the boundary, and not the interior of the reachable set. Our point is that it is an error to look for a uniformly distributed cloud of points in the interior of the reachable set. What we need is the boundary of the set than can be defined by smaller number of attainable points. One could suppose that to assess the shape of the reachable set, we can generate a number of trajectories that belong to its interior, and see the boundary of obtained cloud of points. However, this is not true. Such *simple random shooting* gives wrong results, very different from the true shape of the reachable set.

Shortly speaking, the solver algorithm uses some results from the optimal control theory (Pontryagin [22]). From Pontryagin's principle of maximum, it is known that each model trajectory that reaches a point on the boundary of the reachable set at the final simulation time must entirely belong to the boundary of this set for all earlier time instants. Moreover, such trajectory must satisfy the Jacobi–Hamilton equations (the necessary condition). These equations involve a vector of auxiliary variables $p = (p_1, p_2, p_3, ..., p_n)$.

The algorithm generates a series of boundary-scanning trajectories with randomly generated initial conditions $p(0)$. These trajectories obey the equations of Hamilton–Jacobi. After integrating a sufficient number of trajectories, we can see the shape of the reachable set boundary, see Raczynski [23].

## 4.4 Example: The Lotka–Volterra Model

Lotka–Volterra (L-V) equations describe the dynamics of prey–predator ecological systems (Takeuchi [26]). In the simplest case of two species, the prey population (for example, rabbits) grows due to the birth-and-death process. The growth would be exponential, but it is a limitation: there is a predator (e.g., wolves) who eat rabbits. The population of wolves grow when they have food, but if there are few rabbits available, the wolves die. Denote the rabbit population size as $x_1$ and the wolves as $x_2$. The classical form of two-species Lotka–Volterra equations is as follows:

$$\begin{cases} \dfrac{dx_1}{dt} = ax_1 - bx_1x_2 \\[2mm] \dfrac{dx_2}{dt} = -cx_1 + dx_1x_2. \end{cases} \tag{4.8}$$

In the first equation, the term $bx_1x_2$ means that the rate of rabbits caught by wolves is proportional both to wolves' number and rabbits' number. A similar term appears in the second equation, telling that the growth rate of wolves increases when they have more food. Coefficient $a$ defines the rabbits natural grows rate, and $c$ defines the wolves natural death rate. There are many other versions of the equations used in ecological models, with two or more species in the N-species food chains.

A simple simulation of the above equation is shown in Fig. 4.1.

Now, consider the two-species system with some uncertainty. Namely, suppose that the growth rate of the rabbits is uncertain, subject to climate changes and other external factors. For example, suppose that the parameter $a$ may change in time, within the range of $\pm 25\%$. Thus, the system equations can be written as follows:

$$\begin{cases} \dfrac{dx_1}{dt} = a(1+u)x_1 - bx_1x_2 \\[2mm] \dfrac{dx_2}{dt} = -cx_1 + dx_1x_2, \end{cases} \tag{4.9}$$

where $u$ changes between $-0.25$ and $+0.25$. This way, we obtain a differential inclusion, with the right-hand side parametrized by the variable u.

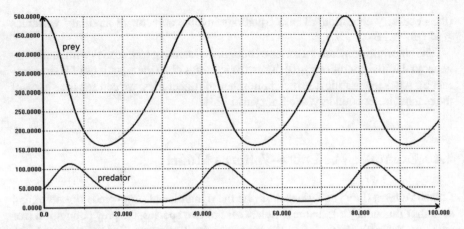

**Fig. 4.1**  A simple simulation of the Lotka–Volterra model

**Fig. 4.2**  The time section of
the reachable set for the L-V
equations. Time $= 45$

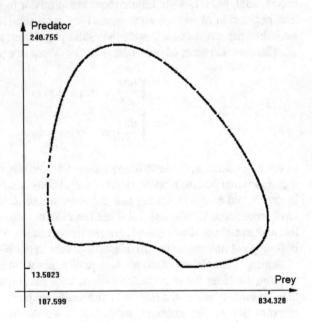

The L-V equations have strong non-linearities (products of two state variables).
Thus, the solution reveals a series of non-sinusoidal oscillations. The model parameters
are $a = 0.1$, $b = 0.02$, $c = 0.3$, $d = 0.001$ and final simulation time equal to 100.

Now, applying the DI solver we obtain the attainable set of the size of the two
species in presence of uncertainty. Figure 4.2 shows the shape of the reachable set
boundary for time $= 45$. In Fig. 4.3, we can see the 3D image of the set.

In ecology and population growth models, we almost always have uncertain factors
that have unknown probability distribution and other probabilistic properties. In these
cases, the differential inclusions may be a useful research tool. Observe that for the

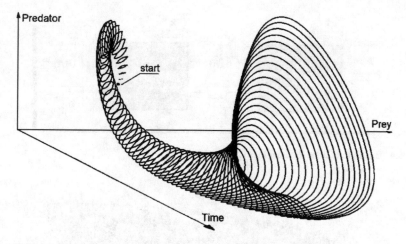

**Fig. 4.3** The shape of the reachable set of the L-V equation. 3D image

Lotka–Volterra model, even with small fluctuations of uncertain parameter, the size of the reachable set after time approximately equal to the model oscillation period is quite big. This means that this model is hardly useful for the predictions, even for small time intervals.

## 4.5 A Mechanical System

A simple mechanical system is shown in Fig. 4.4. The system equations are as follows:

$$\begin{cases} x_1' = x_3 \\ x_2' = x_4 \\ x_3' = \frac{1}{M_1} [F - K_1(x_1 - x_2) - B_1 f(x_3 - x_4)] \\ x_4' = \frac{1}{M_2} [K_1(x_1 - x_2) + B_1 f(x_3 - x_4) - K_2 x_2 - B_2 f(x_4)]. \end{cases} \tag{4.10}$$

Note that the dumpers may be non-linear. $F$ is an external force that belongs to $[-0.5 , 0.5]$. The time section of the reachable set for this system is shown in Fig. 4.5. Here, $time = 4, m_1 = 1, m_2 = 2, k_1 = 0.3, k_2 = 0.1, b_1 = 1.5, b_2 = 3.0$. Some points appear to belong to the interior of the set. Those are not the points

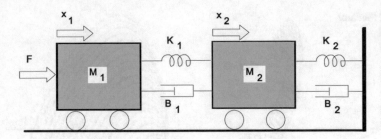

**Fig. 4.4**   Example of a mechanical model

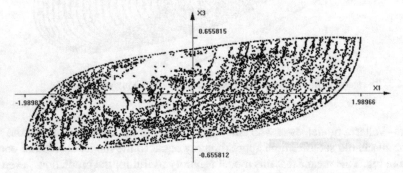

**Fig. 4.5**   The shape of a time section of the reachable set of a fourth-order system of Fig. 4.4

obtained by primitive shooting. In fact, all visible points belong to the boundary of the reachable set. What we see is only a projection of a four-dimensional figure (point cloud) onto a two-dimensional plane $x_1$, $x_2$ for a given time instant (Fig. 4.6).

The graphical representation of the reachable sets for models of dimensionality greater than 3 is somewhat difficult. The question is how to display a cloud of points of n-dimensional space in order to show clearly the shape. If the cloud is three dimensional, this can be done by displaying a rotating 3D image to produce an illusion of 3D viewing. Other possible enhancements may be obtained using techniques known in fuzzy set theory. Figure 4.6 shows the result of calculating the fuzzy variable representing the level of membership in the region. Points with membership value greater than 0.5 are shown as gray pixels. If there are not enough points to analyze, then the holes in the region may appear. Anyway, such images always depict approximate shapes.

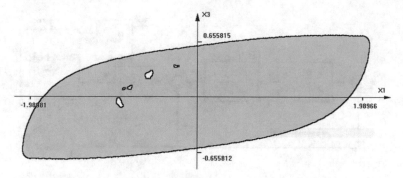

**Fig. 4.6** The shape of the reachable set of Fig. 4.5 enhanced by fuzzy sets technique

## 4.6 Functional Sensitivity of the V/f Speed Control of Induction Motor

The angular velocity $w$ of the motor is given by the following equation:

$$x' = \frac{T - L}{I},$$
(4.11)

(prime sign stands for time differentiation) where $T$ is the torque (Nm), $L$ is the mechanical load (Nm), and $I$ is the moment of inertia of the rotor (kg m$^2$). The formula for $T$ used in this paper is as follows:

$$T(V, s) = \frac{kV^2 r}{\left(z_1 + \dfrac{r_2}{s}\right)^2 sn_s}, \quad k = \frac{f_n}{2\pi},$$
(4.12)

where

$T$—torque ($Nm$),

$f_n$—nominal frequency,

$V$ = supply voltage/1.73,

$z_1$—stator impedance,

$r_2$—rotor resistance,

$n_s$—synchronous velocity r/min (supposed equal to 1800 for $f$ = 60 Hz in the following),

$s = (ns\text{-}n)/ns$ (the slip),

$n$—motor velocity r/min,

$P$—number of pole pairs (supposed equal to 4 in the following)

.

Unlike the DC motor, the velocity of the induction motor cannot be easily controlled by changing the supply voltage or stator current. The three-phase source voltage inverters (VSI) must be used. These devices are based on bipolar transistor

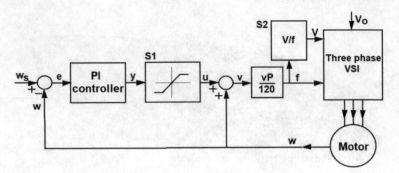

**Fig. 4.7**   Block diagram of a V/f speed control

(IGBT) semiconductor switches. There are other alternatives to the IGBT: insulated gate-commutated thyristors (IGCTs) and injection-enhanced gate transistors (IEGTs). Here, we do not discuss the technology of the VSIs. We assume an ideal VSI that provides the three-phase supply of voltage $V$ and frequency $f$, receiving a necessary power supply. For more information about the VSI control consult Mujal-Rosas and Orrit-Prat [19] or Chen et al. [7].

Changing the frequency, we can change the velocity of the motor. However, such control with constant voltage makes the stator current grow for low frequencies and produce saturation of the air gap flux. Therefore, the stator voltage should be reduced according to the frequency, to maintain the air gap flux constant. The magnitude of the stator flux is proportional to the ratio of the stator voltage and the frequency. Hence, if the ratio of voltage to frequency is kept constant, the flux remains constant. This method is referred to as V/f speed control. Figure 4.7 shows a typical scheme of a V/f control, with a PI controller.

Here, we have:

$w_s$—the set point (desired velocity, rpm)

$$e = w_s - w \text{ (control error)} \tag{4.13}$$

$$y = K\left(e + \frac{z}{T_i}\right) \tag{4.14}$$

$$z(t) = \int_0^t e(\tau)d\tau \tag{4.15}$$

$$u = \begin{cases} y \ if \ |y| \le y_m \\ -y_m \ if \ y < -y_m \\ y_m \ if \ y > -y_m \end{cases} \tag{4.16}$$

$$v = u + w \tag{4.17}$$

$$f = \frac{vP}{120} \tag{4.18}$$

$$V = V_{r}ef\frac{f}{60} \tag{4.19}$$

$$\frac{dw}{dt} = \frac{T(V,s) - L}{I}, \tag{4.20}$$

where $I$ is the moment of inertia of the rotor (kgm$^2$) and $L$ is the load (Nm).

Here, we assume $V_{ref} = 440/\sqrt{3} = 254.03$. The velocities $ws, w, v$ and $u$ are given in *rpm* units, frequency $f$ in Hz. From Eq. (4.15), we obtain

$$\frac{dz}{dt} = e(t). \tag{4.21}$$

Equations (4.20) and (4.21) form the ordinary differential equation model of the system dynamics, where $w$ and $z$ are the state variables. Block S1 of Fig. 4.7 is a delimiter. Block S2 also includes the necessary saturation restrictions.

To illustrate the action of the PI controller, let us show a simple simulation of the control circuit of Fig. 4.7. The model parameters are as follows:

$K = 0.5, T_i = 2$ s, $y_m = 200, P = 4, V_{ref} = 254$ V, $L = 10.1$ N$_m$, $I = 0.08$ kgm$^2$.

The simulation starts with motor rotating with 1800 rpm (synchronous speed, set point), no load. The load is applied at *time* $= 0$. First, the velocity decreases because of the load. The PI controller makes the velocity go back to 1800 rpm. The presence of the integral part of the controller makes the steady-state velocity equal exactly to the set point. At *time* $= 50$, the set point changes to 1900 rpm. The transient process can be seen in Fig. 4.8.

It should be noted that in this simulation and in the following sensitivity analysis, the controller setting is not optimal. We intentionally set the duplication time $T_i$ to a relatively small value, to make the transient processes oscillatory. The controller gain is rather small, to avoid saturations and possible instability.

Now, let us see the functional sensitivity of the model. As stated before, the main topic of this paper is not the model of the control system and simple simulations. Here, we focus at the functional sensitivity. Such analysis can be useful while treating with robust control design. Unlike the conventional sensitivity, the functional sensitivity is a dynamic analysis, based on the differential inclusions. Note that the reachable

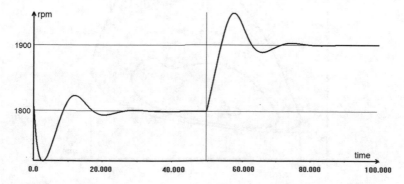

**Fig. 4.8** Simple simulation of the control system of Fig. 4.7

sets we calculate are not obtained by application of simple disturbances. We use the DI solver (Sect. 4.3 and Chap. 3), then scan the boundary of the reachable set using some results of the optimal control theory.

The control system of Fig. 4.7 has two external signals that may be treated as disturbances: the load $L$ and the set point $w_s$. Of course, any other model parameter can be changed. If the variables $L$ and $w_s$ scan the interior of a given permissible set, then the right-hand sides of (4.20) and (4.21) scan certain set $F \subset R^2$. This is the right-hand set of the corresponding differential inclusion. The reachable set of this inclusion defines the functional sensitivity of our model.

Figure 4.9 shows the boundary of the reachable sets with different model parameters. These are intersections of the reachable set with the plane $time = 20$. Contour A was obtained for $L$ and $w$ fluctuating by $\pm5\%$ of their nominal value. Model parameters are as follows:

$K = 0.5, T_i = 2\,\mathrm{s}, y_m = 200, P = 4, V_{ref} = 254\,\mathrm{V}, L = 10.1\,\mathrm{Nm}, I = 0.08\,\mathrm{kgm}^2,$ $setpoint = 1800\,\mathrm{rpm}.$

Contour B was obtained with the fluctuations $\pm8\%$. The model is quite sensitive with respect to the fluctuations of $w$ . If we fix $w$, supposing the load $L$ fluctuating by $\pm5\%$, the reachable set boundary is given by the contour C. This small sensitivity with respect to $L$ is because of a relatively small value of $L$, and due to the action of the controller.

Model parameters for sets of Figs. 4.9 and 4.10 correspond to the stable operating region. If we change the saturation restriction and make the controller action stronger, then the model non-linearities reflect in the images, and the sensitivity

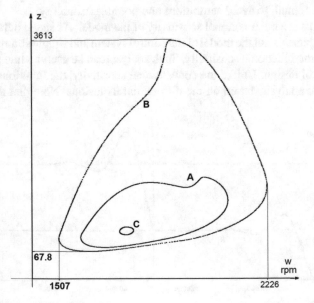

**Fig. 4.9**  System reachable sets for time $= 20$

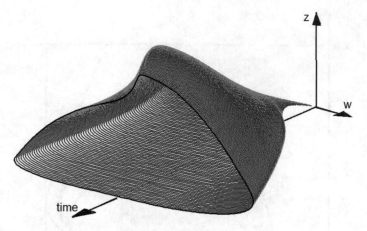

**Fig. 4.10** The 3D image of the functional sensitivity set

grows. This may affect the robustness of the system. Figure 4.11 shows the contour of the reachable set for the following parameters:

$K = 1.6, Ti = 2\,\text{s}, y_m = 700, P = 4, V_{ref} = 254\,\text{V}, L = 10.1\,\text{Nm}, I = 0.08\,\text{kgm}^2,$
$setpoint = 1800\,\text{rpm}, L\ and\ w\ fluctuation\ 8\%.$

The reachable set becomes "irregular," approaching the break-down torque and unstable operating region. In such cases, the DI solver algorithm may reveal some irregularities, caused mainly by the time discretization and imperfections of the numerical integration algorithm.

### 4.6.1 Comparison with the Classical Sensitivity Analysis

Let us run the model with another saturation restriction and greater final time, as follows:

$K = 0.5, T_i = 2\,\text{s}, y_m = 700, P = 4, V_{ref} = 254\,\text{V}, L = 10.1\,\text{Nm}, I = 0.08\,\text{kgm}^2,$
set point $= 1800$ rpm, final time $= 50$.

Figure 4.12 depicts the contour of the final reachable set. Note the set of points marked with X. The solver displays these points as the boundary points, but, in fact, they belong to the interior of the reachable set. This is due to the fact that the solver is based on the concepts of the maximum principle that provides necessary and not sufficient conditions for optimality. Anyway, each displayed point belongs to the reachable set.

In Fig. 4.12, we also can see the reachable set obtained by the "Vensim-like" sensitivity analysis. Analysis of this kind is provided by many System Dynamics software tool, like Vensim, Powersim, and others. In this type of analysis, a series of trajectories with randomly changed model parameters ($w$ and $L$ in our case) is calculated. The parameters are different for each trajectory, but constant along with

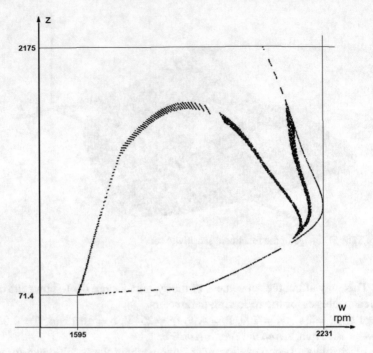

**Fig. 4.11** Strong non-linearities revealed with greater amplitudes of the variations

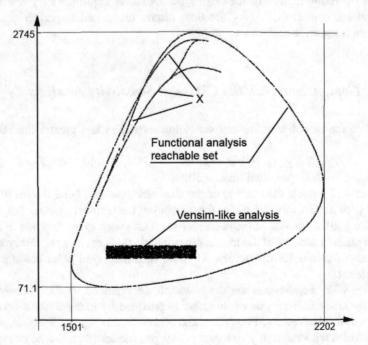

**Fig. 4.12** Reachable set for time = 50 and Vensim-like sensitivity set

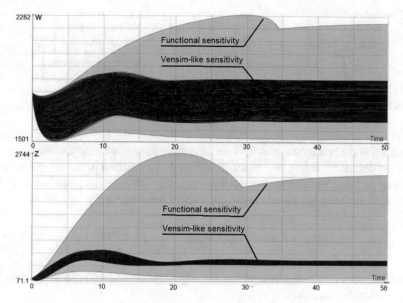

**Fig. 4.13** Reachable set projections to (time-w) and (time-z) planes

each trajectory. It can be seen how different sensitivity sets are provided by the functional sensitivity. Figure 4.13 shows the projections of the true reachable set and the Vensim analysis into the time-state plane (a "side view").

## 4.7 PID Anti-Windup Control

Functional sensitivity can be an important tool in robust control system design. Such systems should not only be stable and perhaps optimal, but also must work satisfactory with adverse external disturbances. For more information on robust control consult, for example, Xu and Lam [27] or Zhou and Doyle [30].

Figure 4.14 shows a closed-loop control system. The controller is of the proportional–derivative–integral action, and the process is of second order, oscillatory. The set point is denoted as $p$, $e$ is the control error, $z$ is a disturbance, $q$ and $x$ are the process input and output, respectively. The control error $e = p - x$. Two versions of the controller are used: the classic PID and the anti-windup version. At the controller output, there is an actuator that may saturate when its limit $L$ is reached. The classic PID is described by the following equation (in terms of the Laplace transform).

$$y(s) = K_R e(s)\left(1 + T_d s + \frac{1}{T_i s}\right),$$

(4.22)

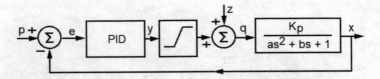

**Fig. 4.14** PID control circuit with disturbance

where $K_R$ is the controller gain, $T_i$ is the integrator parameter (duplication time), and $T_d$ is the derivative action parameter. To avoid differentiation of the set-point signal that may have discontinuities, we use the controller where the set point is not applied to the differentiator (a special case of the "set-point weighing").

$$y(t) = K_R \left( e(t) - T_d \frac{dx}{dt} x + \frac{1}{T_i} \int_0^t e(\tau) d\tau \right), \tag{4.23}$$

where the error $e = p - x$ is replaced by $-\frac{dx}{dt}$. The controlled process is described by the following equation:

$$a \frac{d^2 x}{dt^2} + b \frac{dx}{dt} + x(t) = K_R q. \tag{4.24}$$

Further analysis of the system requires the canonical (vectorial) form of the model equations. To obtain this for equations, first let us denote

$$g(t) = \frac{1}{T_i} \int_0^t e(\tau) d\tau. \tag{4.25}$$

Now, we define the state vector $(x_1, x_2, x_3)$, so that

$$x_1 = x, \ x_2 = \frac{dx}{dt}, \ x_3 = g. \tag{4.26}$$

From (4.23), (4.24), and (4.25), we obtain

$$\begin{cases} \dfrac{dx_1}{dt} = x_2 \\[2mm] \dfrac{dx_2}{dt} = (K_p(y + z) - bx_2 - x_1)/a \\[2mm] \dfrac{dx_3}{dt} = \dfrac{v}{T_i}(p - x_1) \\[2mm] \dfrac{dx_4}{dt} = \dfrac{dg}{dt} = K_R(e - T_d x_2 + x_3). \end{cases} \tag{4.27}$$

The variable $v$ is added to simulate the anti-windup PID controller. This PID version is implemented to avoid the windup effect, when the value of the integration action may become very large. This occurs when the control error is positive or negative during long time interval. In practical applications, due to the windup, the actuator limit may be reached and the control signal $y$ becomes constant.

There are several analog versions of the anti-windup controller (consult Astrom and Hagglund [3]). However, in newer PID implementation, the anti-windup is done by software. It turns off the integration action when the saturation occurs. The variable $v$ is defined as follows:

$$v(e, y, L) = \begin{cases} 0 \text{ if } |y| \geq L \\ 1 \text{ if } |y| < L \text{ or } (y \geq L \text{ and } e < 0) \text{ or } (y \leq -L \text{ and } e > 0), \end{cases}$$

(4.28)

where $L$ is the saturation level. Variable $v$ shuts down the integrator action when the saturation level is reached and turns it ON when the integrated value is equal or greater than $L$, and $e$ is negative, or when the integrated value is equal or lower than $-L$ and $e$ is positive.

As stated before, the functional sensitivity is given by the reachable set of the model.

Consider the model (4.27) and (4.28).

The parameters are as follows:

$a = 2, b = 1$ (process parameters),

$p = 1$ (the set point),

$K_R = 2.5$ (controller gain),

$T_i = 2$ (duplication time),

$T_d = 0.2$ (derivative time),

$L = 1.6$ (actuator saturation level).

The above PID settings are not optimal. They were defined this way, to exaggerate some imperfections in the transient process. Anyway, a robust control system should work not only with optimal parameters.

Suppose that both the values of the disturbance $z$ and the process gain $K_R$ are uncertain and may fluctuate by $\pm 20\%$ around their original values. The system response reaches a steady state after about 30 time units. However, we are interested rather in system behavior during the transient process, so we will investigate the functional sensitivity at shorter time, $t = 8$. Figure 4.15 shows the reachable set of the process response to unit step input, process only, open loop. In Fig. 4.16, we can see some, randomly selected (but not random), trajectories that scan the boundary of the reachable set for process only.

Now, let the same parameter fluctuations be applied in the closed-loop control circuit. First, we simulate the system without actuator saturation. The closed-loop model is of order three. Thus, the reachable set boundary at a given final time instant is not just a contour, but a three-dimensional cloud of points. The points are located on a three-dimensional surface, like a surface of a balloon. Consequently, we cannot get a two-dimensional contour. We can only see projections of the reachable set boundary

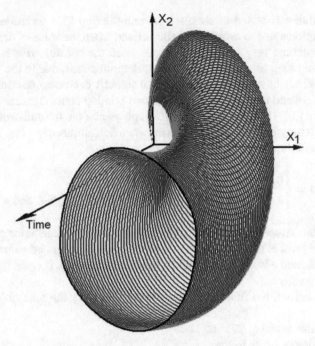

**Fig. 4.15**  Open-loop reachable set for the process only. 3D image generated by the DI solver

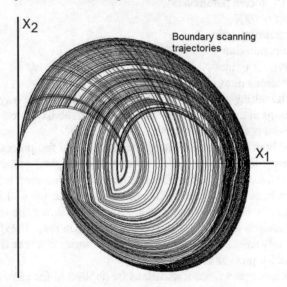

**Fig. 4.16**  Some boundary-scanning trajectories, process in open loop

into a two-dimensional plane. The DI solver generates a lot of such projections. Let us show only some examples, namely, the projections into the plane $x_1 - x_3$ and $time - x_1$.

Figure 4.17 shows the projection of the reachable set at $t = 8$ into the plane $x_1 - x_3$.

Figure 4.18 depicts the "side view" of the same reachable set, projected into the $time - x_1$ plane. Now, consider the anti-windup controller with actuator with saturation level equal to 1.6. Figure 4.20 shows, like Fig. 4.17, the projection of the boundary points at the $x_1 - x_3$ plane. Part B shows also the points obtained by the classical risk analysis, with the same range of parameter changes. It can be observed the difference in the shape of the reachable set. Note the difference between the maximal value of x1 (overshoot) in Figs. 4.17 and 4.19, equal to 1.59 and 1.31,

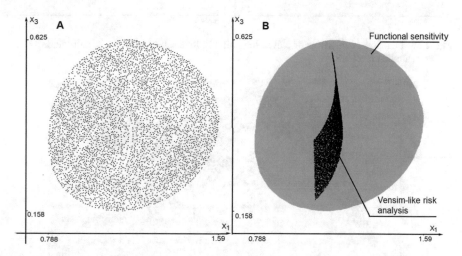

**Fig. 4.17** Functional sensitivity of closed-loop circuit. Reachable set at t = 8, no saturation

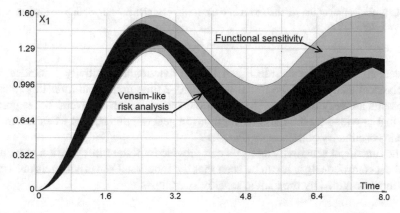

**Fig. 4.18** Projection of the reachable set into the $time - x_1$ plane

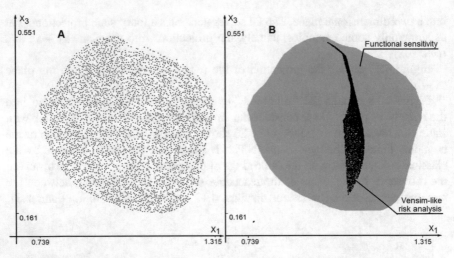

**Fig. 4.19** Functional sensitivity of closed-loop circuit. Reachable set at $t = 8$, anti-windup controller, saturation level 1.6

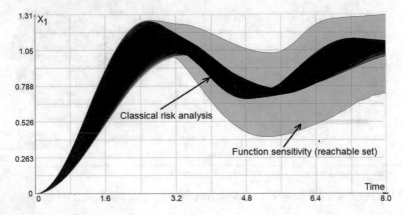

**Fig. 4.20** Projection of the reachable set into the $time - x_1$ plane. Anti-windup controller, saturation level equal to 1.6

respectively. As can be seen in Fig. 4.18, the functional sensitivity coincides with the classical analysis for the initial interval (0–2 time units), but after this interval the results are very different. The functional sensitivity set is several times greater than that provided by the classical method. In general, this occurs when the model contains oscillatory components (Fig. 4.19).

The functional sensitivity analysis can provide important information to be used in robust control design. It can be seen that the functional sensitivity sets are larger than those provided by the classic sensitivity analysis.

## 4.8 Vehicle Horizontal Movement

As mentioned before, the topic of this chapter is not the model design and validation, but the functional sensitivity and reachable sets of some known models. The model used below is used to show the functional sensitivity of the car horizontal movement (Fig. 4.21). The results may help to design a robust control system for autonomous car control system.

In Zhang [29], we can find the phase plane and stability analysis of the model. Pepy et al. [21] use this model to vehicle path planning. Here, we use the model with two degree of freedom, supposing the longitudinal car velocity to be constant.

Let us denote

$m$—vehicle mass,

$v_y$—lateral velocity,

$\Psi$—vehicle path angle (in absolute coordinates),

$r$—yaw velocity $d\psi/dt$,

$\delta$—front wheel angle,

$F_{sf}$—front wheel lateral force,

$F_{sr}$—rear wheel lateral force,

$L_f$—distance between front axle and center,

$L_r$—distance between rear axle and center,

$I$—vehicle yaw moment of inertia,

Radians and SI unit system are used.

The model equations, due to Zhang [29], are as follows:

$$\begin{cases} \dfrac{dv_y}{dt} = \dfrac{F_{sf} \cos \delta + F_{sr}}{m} \\ \dfrac{dr}{dt} = \dfrac{F_{sf} L_f \cos \delta + F_{sr}}{I}. \end{cases} \tag{4.29}$$

**Fig. 4.21** Car horizontal movement

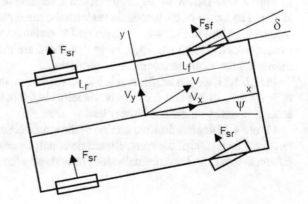

The wheel lateral forces depend on the following angles:

$$\begin{cases} \alpha_f = \delta - arctan\left(\dfrac{v_y + rL_f}{v_x}\right) \\[2ex] \alpha_r = -arctan\left(\dfrac{v_y - rL_r}{v_y}\right). \end{cases} \tag{4.30}$$

The formula for the lateral forces is strongly non-linear

$$F = D\,sin\,(C\,arctan(B\alpha - E(B\alpha - arctan B\alpha))), \tag{4.31}$$

where $F$, $D$, $B$, $C$, and $\alpha$ have suffix $sf$ or $rf$ for front and rear wheel, respectively.

The values of model parameters used in the simulations are as follows:
$m = 1640$ kg, $v_x = 20$ m/s, $I = 2900$ kgm2, $L_f = 1.1$ m and $L_r = 1.4$ m.
The parameters of the lateral force formula are
$B_f = 11.27$, $C_f = 1.56$, $D_f = 2575$, $E_f = -1.999$
$B_r = 19.53$, $C_r = 1.56$, $D_r = 1750$, $E_r = -1.79$.

The simulations and phase portraits of model (4.29) are given in Zhang [29]. Here, we will calculate the functional sensitivity of the model determining the corresponding reachable sets. Recall that in this kind of analysis we assume that some model parameters have uncertain values that may change in time. The parameters that are subject to such changes are mainly those related to the wheel lateral forces. While the vehicle mass and moment of inertia can hardly change in time, the lateral forces do. This may occur due to the irregularities of the road, obstacles, and variable friction coefficients.

Thus, let us suppose that the lateral forces $F_{sf}$ and $F_s r$ have uncertain values and may change in time by $\pm 10\%$ of the value calculated by Formula (4.31). We will analyze the model for different steering angles. Suppose that the steering angle $\delta$ is equal to 0.04 radians (2.29 degree), and the model final time is equal to 1.5 s.

Figure 4.22 shows the 3d image of the reachable set viewed from two different angles. The image of the reachable set with the same parameters and $\delta$ equal to 0.25 radians is shown in Fig. 4.23. The range for variable vy is $(-5.09, -0.57)$ and for r (yaw velocity) is $(0.10, 0.52)$. Note that those are the model variables in the car frame, not the absolute velocities or positions.

In Fig. 4.23, we can see the reachable set for $\delta$ equal to 0.25. The range for variable $v_y$ is $(-4.35, -0.27)$ and for $r$ (yaw velocity) is $(-0.009, 0.37)$. The shape changes according to the model non-linearities.

Figure 4.24 depicts the time section of the reachable set of Fig. 4.23, for time equal to 1.5 s. The results of the conventional risk analysis are shown as the black cloud of points. In this case, the extremal values provided by the functional sensitivity and by

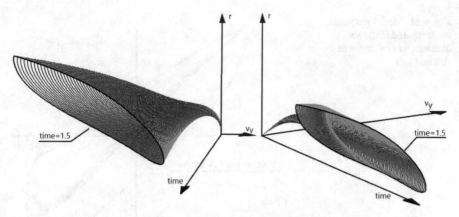

**Fig. 4.22**  3D image of the reachable of model (4.29). Two different view angles

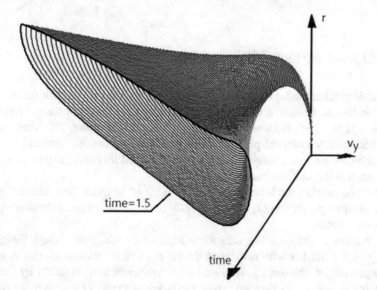

**Fig. 4.23**  The reachable set of model (4.29) with $\delta$ equal to 0.25

the Vensim-like analysis are similar. This occurs when the model is more dissipative or well dumped. As can be seen from the previous models, in case of oscillatory system, the discrepancy between the sensitivity RS and the conventional analysis may be rather great. Anyway, the set provided by the conventional analysis (4.24) is significantly smaller than the true reachable set.

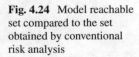

**Fig. 4.24** Model reachable set compared to the set obtained by conventional risk analysis

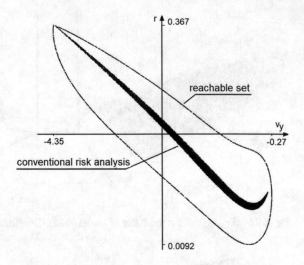

## 4.9   Marketing Sensibility and Reachable Sets

A dynamic market model is used, with uncertain parameters. We look for reachable regions in the model state space for possible strategies in the investment. The control parameter is the share of investment, being a part of the revenue. This share, as well as the uncertainty in model parameters, is treated using the differential inclusions. The reachable sets are obtained, using the differential inclusion solver (Sect. 4.3). Other result is the optimal control strategy.

The same market model is used as the object of optimization where Pontryagin's maximum principle [22] is used directly to obtain some more detailed optimal marketing policies.

The market model itself is not the main topic of this chapter. Though the optimal strategy of the market with respect to the investment is calculated, this is not the main topic, either. The work is focused on the treatment of uncertainty by means of the reachable sets, rather than stochastic modeling. The model we are using (at least the model of the demand) is taken from the literature, with some dynamic properties added.

A lot of demand models may be used for similar purpose. A comprehensive overview can be found in Lilien and Kotler [15] (linear, non-linear, deterministic, stochastic, static, dynamic, etc.). What we need is a model that can be adopted into dynamic market simulation. The demand model we choose is taken from that book. The dynamics is added to simulate the inertia of total accumulated revenue, and the market growth due to the investment. The growth is charged with certain inertia, which implies additional differential equations (see the next section for details).

The search for market models for simulation purpose dates from early 1960s, and significant publications began to appear in late 60s. King [14] gives a comparison of iconic, analog, and symbolic models. Recall that an iconic model represents reality on a smaller scale, an analog model shows reality by means of maps and diagrams, and a symbolic model uses mathematical expressions to portray reality. Montgomery et al. [16] discuss the descriptive, predictive, and normative models, where by descriptive we mean models that consist largely of diagrams and maps or charts designed to describe a real-world system. Predictive model is used in predictive analytics to create a statistical model of future behavior, and normative model evaluates alternative solutions to answer the question, "What is going on?" and suggests what ought to be done or how things should work according to an assumption of standard.

Optimal control in marketing is not a new topic. There are many publications in the field like, for example, Bertsimas and Lo [4]. In that article we can find the application of the Bellman's dynamic programming approach to the problem of the price impact on the dynamic trading on the stock exchange. A tutorial and survey on the relevant technical literature on models of economic growth can be found in Burmeister [5]. Feichtinger et al. [10] consider a similar problem for a general market and optimal advertising policy. In that article a detailed formulation of an implementation of the Pontryagin's Maximum Principle is shown. Yuanguo Zhu [28] uses the Bellman's Principle of Optimality, and derive the principle of optimality for fuzzy optimal control. This is applied to a portfolio selection model. As stated earlier, the optimization is not our main subject. However, the presented method can also be used to obtain optimal investment strategy. Anyway, to determine reachable sets we use methods of the optimal control theory.

### 4.9.1  The Model

The demand model we use is a non-linear model of the demand and revenue which parameters are: price, advertising, seasonal index, market overall growth, consumer income, production or acquisition cost and the market elasticities with respect to the price, advertising and consumer income (Lilien and Kotler [15]). The demand model is as follows.

$$q(p, a, y, v, g, t) = q_0 v(t) s(t) \left( \frac{p(t)}{p(0)} \right)^{e_p} \left( \frac{a(t)}{a(0)v(t)s(t)} \right)^{e_a} \left( \frac{y(t)}{y(0)} \right)^{e_y}, \quad (4.32)$$

where $q$—demand, $q_0$—initial or reference demand,

$t$—the time,

$s(t)$—overall market size,

$p(t)$—price,

$a(t)$—advertising per time unit,

$y(t)$—consumer income,

$v(t)$—seasonal index,

$e_p, e_a, e_y$—market elasticity with respect to price, advertising, and consumer income, respectively.

In the model of Lilien and Kotler, the market size $s(t)$ is supposed to be equal to $(1 + g)^t$, where $g$ is the market growth rate. We use the variable s instead, to be able to link the market growth to the investment that the company may do in order to expand the market (new installations, infrastructure, etc.).

Normally, the price elasticity is negative (price increase means less demand), and the elasticity for advertising and consumer income are positive. Note that the term $v(t)s(t)$ multiplies the demand, and appears also in the denominator of the advertising impact term. This means that if the market and the seasonal index grow, then we must spend more on advertising to achieve the same effect. We do not consider any storage or warehouse mechanism, so it is supposed that the sales are equal to the demand. The revenue can be calculated as follows:

$$U = (I - b)e - a, \text{ where } e = (p - c)q, \tag{4.33}$$

where $U$ is the net revenue, $c$ is the unit cost of the product (production, acquisition cost), $a$ is the advertising, $b$ is the investment factor, $p$ is the price to the public, and $Q$ is given by formula (4.32). The investment b determines what part of the utility e is being invested in the market growth, $0 \leq b \leq 1$. The market growth rate is proportional to the product $be$. It will be one of our control variables. In our equations, the market size is relative with initial condition equal to 1. If the company does not invest in the market ($b = 0$), then the market size has no influence on the model trajectory (remains constant). The final *total revenue* is the sum (integral) of $U$ over the time interval under consideration.

Figure 4.25 shows the shape of the profit (utility) e as a function of the price and advertising, when the time and other variables are fixed.

It can *be* seen that the utility function has a maximum that determines the optimal price and advertising at the moment. This is a static case. Now, consider a fixed time interval $[0,T]$. To calculate the total revenue, we must integrate (4.33) over the time. The variable $b$ controls the investment, $0 < b < 1$. The market size growth due to the investment. To make the model more realistic, we suppose that the market growth rate follows the investment with certain inertia. The basic equations are as follows:

$$\begin{cases} \dfrac{dr}{dt} = (1 - b)v/v_0 \text{ accumulated revenue growth, relative} \\ v = (p - c)q(z, w, y, v, s, t) - a_0, \quad \text{—initial value} \\ \dfrac{ds}{dt} = (x - s)/T_c, \text{ market size growth, relative} \\ \dfrac{dx}{dt} = \dfrac{gbv}{v_0}, \text{ accumulated investment, relative} \end{cases} \tag{4.34}$$

In the above equations, $r$ is the relative accumulated net revenue, $s$ is the relative market size, and x is an auxiliary variable. $T_c$ is the time constant of the market growth inertia and $g$ is a constant that defines the impact of the investment on the market expansion. Namely, $g$ tells what is the relative market growth per one currency unit

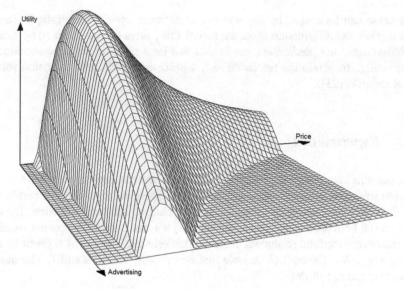

**Fig. 4.25** The revenue as function of the price and advertising. Static case, b = 0

invested. All other variables may change in time. Observe that the additional state variable $x$ has been introduced to manage the inertia.

Note that our model (4.34) has the form of a control system and defines the corresponding DI. In the investment strategy problem, the control variable is $b(t)$, the share of investment we apply out of the total revenue (experiment 1, below). The uncertainty of the other model variables can be treated in a similar way, letting the variable change within a given interval. Below are presented the results of such analysis for the uncertainty in the values of $e_p$—market elasticity with respect to the price.

The images of the sensitivity reachable sets shown in the following experiments have been generated by the differential inclusion solver, described in Sect. 4.3 and Chap. 3.

The final accumulated revenue and the market size strongly depend on the changes of the investment control variable $b$. So, it is important to be able to see the limits of the revenue and the market size when $b$ changes due to the investment strategy. Note that if $b = 1$ all the time (maximal investment), then we have no revenue available, and if $b = 0$ then the market does not grow, which also reduces the available profit.

The aim of the present work is to calculate the reachable sets for all possible investment strategies, as well as to assess the impact of the uncertainty of important market variables.

This should be emphasized that our approach to uncertainty has nothing to do with randomness or stochastic models. If a value of a parameter is uncertain, this means that it can take values from some interval, but this does not mean that it is a random variable. Such a parameter does not have any probability distribution. The changes in its value can be caused by any internal or external agents. For example, in the stock market, the information about the actual share price can be obtained by means of observations and predictions, but it also can be a false information introduced intentionally. To obtain the reachable sets, we use the differential inclusion solver (see Raczynski [23]).

## 4.9.2  Experiment 1

Now, our aim is to obtain the reachable set in the revenue/market size plane for all possible strategies of investment, with final time fixed. As an additional result, the DI solver provides the control function (investment as a function of time) for any point on the boundary of the reachable set. So, we can obtain the optimal strategy that maximizes the total revenue or the final market size. The model is given by Eq. (4.34), where $b$ is the control variable that can change within 0 and 1. The model parameters are as follows:

Initial sales $= 70000$,
Initial (reference) advertisement $= 10000$ per time unit (a day),
Product unitary cost $= 0.8$,
Initial relative market size $= 1$,
Unit price $= 1$,
Unit initial relative price $= 1.2$,
Time constant for investment inertia $T_c = 30$ days,
$g = 0.006$,
Market elasticity for the product price $= -2$,
Market elasticity for advertisement impact $= 0.6$,
Final simulation time $= 365$ days.

Note that the initial relative price is set equal to 1.2 and not 1. This is to make the demand change when the price elasticity changes. In this experiment, both seasonal index and the consumer income are functions of time. The relative consumer income is equal to 1 except the interval (200, 214) where it is equal to 2.5. The seasonal index is equal to 3 within the interval (100, 128), equal to 0.5 in (160, 167) and equal to 1 elsewhere.

Figure 4.26 shows the image of the reachable set for the revenue and market size at the simulation final time, with the investment control b included in [0, 1]. It should be noted that in our case the dimensionality of the state space is equal to 3. So, to show the complete reachable set we should display it in the four-dimensional time-state space, which is quite difficult. If we limit the displayed state-space components to market size revenue, then the reachable set boundary is not always seen just one contour, because for each time instant the two-dimensional image is a projection of a three-dimensional reachable region. Fortunately, in this case, the boundary is clearly

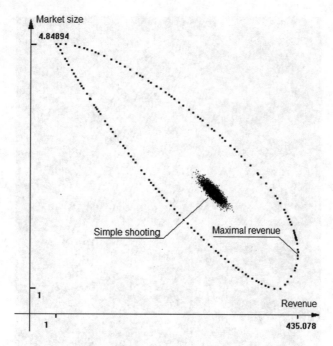

**Fig. 4.26**  Reachable set for experiment 1

seen, perhaps because the control vector dimensionality is equal to 1. On the same figure, we also can see a set of reachable points obtained by simple random shooting (10000 trajectories integrated). It is clear that the simple shooting provides a wrong assessment of the reachable set. Figure 4.27 depicts the 3D image of the reachable set in the time-state space.

This and other simulations show that both the net revenue and the market size may reach greater values than that obtained with fixed elasticity $e_p$.

### 4.9.3  Experiment 2

Now, suppose that the value of the market elasticity with respect to the price $e_p$ is uncertain and may change up to plus minus 20% of its original value. Using model parameters as in experiment 1, with variable investment control in [0,1] and uncertain (variable) $e_p$, we obtain a larger reachable set. In this case, the set is more complicated (2D projection of a 3D region). This results in a cloud of reachable points and not in a simple contour. A post-processing of this cloud provides an assessment of the reachable set, shown as the gray region on Fig. 4.28. To compare with experiment 1, the boundary points of the previous reachable set are also shown on the same figure.

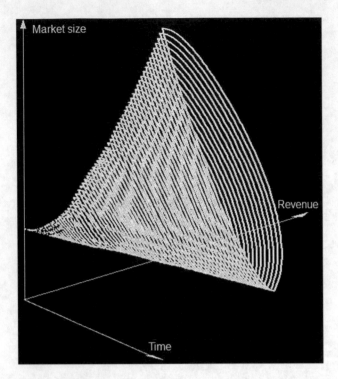

**Fig. 4.27** Experiment 1. 3D image of the reachable set

Both the net revenue and the market size may reach greater values than that obtained with fixed elasticity $e_p$.

Calculating reachable sets is something more than obtaining the optimal control. Though we can get the optimal control strategy as an additional result, the possibility of viewing the shape of reachable sets provides much more information about the model behavior. The same simulation program can be used for many more experiments, looking for reachable sets with variable advertising and uncertainty of any one of the model parameters. Adding more inertia to the model will result with higher state vector dimensionality and more complicated images of the reachable sets. For such sets, the problem is not only the set determination, but also the way to display it. As the computer screen is (still) two dimensional, the good effects can be obtained while displaying the cloud of points in movement, rotating around one of the state-space axes. Unfortunately, such moving displays can hardly be shown as figures of a printed version.

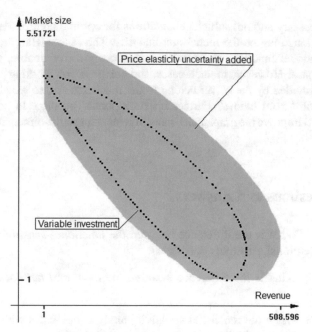

**Fig. 4.28** Reachable set with variable investment and uncertain price elasticity

## 4.10 Conclusion

The functional sensitivity is closely related to the differential inclusions. Here, we use the non-local functional sensitivity that is defined as the reachable set of the model with uncertain parameters. This approach is deterministic. We do not treat the disturbances as random variables. Such deterministic approach seems to be more useful in problems of robust control, when we are interested in "worst-case" behavior, rather than in probabilistic model properties. The differential inclusion solver works satisfactory for the presented models. The classical risk analysis may, in some cases, coincide with the functional sensitivity. However, if the model includes oscillatory components, the results are very different, and the functional sensitivity sets may be several times greater than obtained by classical methods. The very essential concept of the DI solver is that it explores the boundary of the reachable set and not its interior. Though some mechanisms of Pontryagin's maximum principle are used, we do not solve any two-point boundary problem. This makes the solver fast. To estimate the shape of a reachable set for a second-order model, we need between 30 s and 2 min. If the non-linearities are strong, the time may be greater. The calculation time grows significantly for higher order models. A non-linear model of order six needs more than half hour of computing time. Some imperfections of the obtained images may appear when the reachable set is complicated and folds several times. This is normal; recall that the maximum principle provides necessary and not sufficient conditions for optimal control. Another cause of imperfections may be the model non-linearity. The Jacobi–Hamilton equations require the model functions to be twice continuously differentiable, which is not always the case. However, in such cases, the solver keeps working. Anyway, all the points provided by the solver belong to the reachable set, so we always obtain the estimation "from below." Further research should be done to accelerate the algorithm and improve the graphical display of results for multi-dimensional models.

## 4.11   Questions and Answers

**Question 3.1** What is the difference between the *functional sensitivity* of models and the conventional sensitivity definition?

**Question 3.2** What is the difference between the *local and non-local functional sensitivity*?

**Question 3.3** When the functional sensitivity analysis coincides with the conventional, Vensim-like sensitivity, and when it does not?

**Answers**

**Answer 3.1** The *functional sensitivity* provides the reachable sets or the model trajectory related to parameters or disturbances that fluctuate in time. It is defined in terms of the *variational calculus*.

**Answer 3.2** The *local and functional sensitivity* is defined in terms of the variational calculus. The *non-local functional sensitivity* provides reachable sets that correspond to the not necessarily small fluctuations of the uncertain parameters or external perturbations.

**Answer 3.3** The functional and conventional sensitivity may provide similar results for models with highly dissipative or dumped elements. However, when the model includes the oscillatory components, the results may be quite different. Also, for the oscillatory models, the two methods may coincide for short time intervals, less than the oscillation period.

## References

1. Aubin JP, Cellina A (1984) Differential inclusions. Springer, Berlin. ISBN/ISSN 978-3-642-69514-8. https://doi.org/10.1007/978-3-642-69512-4
2. Arora JS, Cardoso JB (2012) Variational principle for shape design sensitivity. Aerosp Res Central 30(2):538–547, AIAA. https://doi.org/10.2514/3.10949

3. Astrom KJ, Hagglund T (1995) PID controllers. Instrument Society of America, ISBN: 1-55617-516-7
4. Bertsimas D, Lo AW (1998) Optimal control of execution costs. J Financ Markets 1:1–50
5. Burmeister E, Dobell AR (1972) Guidance and optimal control of free-market economics: a new interpretation. SMC-2(1):9–15
6. Cacuci DG (2003) Sensitivity and uncertainty analysis. Chapman & Hall/CRC, London. ISBN: 1-58488-115-1
7. Chen L, Wang X-, Min Y, Li G, Wang L, Qi J (2020) Modelling and investigating the impact of asynchronous inertia of induction motor on power system frequency response. Int J Electr Power Energy Syst 117. https://doi.org/10.1016/j.ijepes.2019.105708
8. Cook RD, Weisberg S (1982) Criticism and influence analysis in regression. Sociol Methodol 13:313–461
9. Elsgolc LD (2007) Calculus of variations. ISBN: 978-0486457994
10. Feichtinger G, Hartl RF, Sethi SP (1994) Dynamic optimal control models in advertising: recent developments. Manag Sci INFORMS 40(2):195–226
11. Forrester JW (1961) Industrial dynamics. Pegasus Communications, Waltham, MA
12. Freedman DA (2005) Statistical models: theory and practice. Cambridge University Press
13. Friendly M, Dennis D (2005) The early origins and development of the scatterplot. J Hist Behav Sci 41(2):103–130. https://doi.org/10.1002/jhbs.20078
14. King WR (1967) Quantitative analysis for marketing management. McGraw-Hill, New York
15. Lilien GL, Kotler P (1972) Marketing decision making: a model-building approach. Harper & Row, New York. ISBN: 0060440767
16. Montgomery DB, Urban GL (1969) Management science in marketing. Prentice Hall, Englewood Cliffs, NJ
17. Mordukhovich (2005) Sensitivity analysis for generalized variational and hemivariational inequalities. Adv Anal 305–314. https://doi.org/10.1142/9789812701732_0026
18. Morris MD (1991) Factorial sampling plans for preliminary computational experiments. Technometrics 33:161–174. https://doi.org/10.2307/1269043
19. Mujal-Rosas R, Orrit-Prat J (2011) General analysis of the three-phase asynchronous motor with spiral sheet rotor: operation, parameters, and characteristic values. IEEE Trans Ind Electron 58(5):1799–1811. https://doi.org/10.1109/TIE.2010.2051397
20. Nearing J (2010) Mathematical tools for physics. Dover Publications
21. Pepy R, Lambert A, Mounier H et al (2006) Path planning using a dynamic vehicle model. In: Conference paper: 2006 2nd international conference on information & communication technologies, Damascus, Syria, ISBN/ISSN 0-7803-9521-2
22. Pontryagin LS (1962) The mathematical theory of optimal processes. Wiley Interscience, New York
23. Raczynski S (2002) Differential inclusion solver. In: Conference paper: conference paper: international conference on grand challenges for modeling and simulation, The Society for Modeling and Simulation Int., San Antonio TX
24. Sobol I (1993) Sensitivity analysis for non-linear mathematical models. Math Modeling Comput Exp 1:407–414
25. Sriyudthsak K, Uno H, Gunawan R, Shiraishi F (2015) Using dynamic sensitivities to characterize metabolic reaction systems. Math Biosci 269:153–163
26. Takeuchi Y (1996) Global dynamical properties of Lotka-Volterra systems. World Scientific
27. Xu S, Lam J (2006) Robust control and filtering of singular systems. Springer. https://doi.org/10.1007/11375753
28. Yuanguo Z (2009) A fuzzy optimal control model. J Uncertain Syst 3(4):270–279, www.jus.org.uk
29. Zhang H, Li X, Shi S et al (2011) Phase plane analysis for vehicle handling and stability. Int J Comput Intell Syst 4(6):1179–1186. https://doi.org/10.1080/18756891.2011.9727866
30. Zhou K, Doyle JC (1997) Essentials of robust control. Prentice Hall, ISBN: 0-13-525833-2

# Chapter 5
# Attainable Sets in Flight Control

## 5.1 Introduction

In this chapter, we present an example of application of the differential inclusion solver. The main topic is the generation of the reachable sets and not the model itself.

An important problem in aircraft control is the resolution of conflict situations. What we need is the shape of the set of possible aircraft positions after a maneuver. In other words, the information that the probability of an accident was equal to 0.0001 is not very relevant to a victim of an accident. He/she rather wants to know if the accident or collision is possible or not. This information may be important in many other situations. For example, the classical missile-plane (pursuit-evasion) game cannot be won by the missile if the reachable sets of the plane and of the missile positions do not intersect. This information may also be used while considering differential games.

We use the differential inclusion model (see Chap. 3) as the tool for flight attainable sets calculations. This problem is important for flight safety issues. As for the pursuit-evasion games, a similar approach can also be found in works on the game theory. Grigorieva and Ushakov [5] consider the differential game of pursuit-evasion over a fixed time segment. The attainable set is appointed with the help of the stable absorption operator. A more general, variational approach to differential games can be found in Berkovitz [1]. The DIs are used by Solan and Wieille [13] to study the equilibrium payoffs in quitting games. For general problems of the Game Theory, consult, for example, Petrosjan and Zenkevich [9], Isaacs [7], or Fudenberg and Tirole [4].

The applications of the DIs in flight control problems are not new. There are publications on this topic, mainly using the DIs as a part of optimal control algorithm. Though the solver requires multiple solutions of the system state equations and of the conjugated vector trajectory, it is not the same as to solve a series of dynamic optimization tasks. The examples of applications of the DIs in optimal control can be found mainly in the Journal of Guidance, Control, and Dynamics. For example,

© The Author(s), under exclusive license to Springer Nature Switzerland AG 2022
S. Raczynski, *Models for Research and Understanding*, Simulation Foundations,
Methods and Applications, https://doi.org/10.1007/978-3-031-11926-2_5

Seywald [12] describes an optimization algorithm based on the DIs. The similar problem for desensitized optimal control is described in another paper of Seywald [11]. An extensive and detailed report on an application of the DIs in flight control is given by Dutton [2]. In that paper, the problem of the existence of the return-to-launch-site trajectory for a space vehicle is considered. This is an important topic in the abort mission scenarios, related to the reachable set determination. However, finally, it is converted to an optimal control problem and the return trajectory calculation, namely, the feasible aborts along the ascent trajectory. Other applications of the DIs to optimal control can be found in Mordukhovich [8] Raivio [10], and Fahroo and Ross [3]. However, these problems are quite different from the application described here. What we are looking for is the shape of the whole reachable set and not a particular optimal trajectory. Thus, we need a DI solver as described in Chap. 3, rather than an optimization algorithm.

## 5.2 Control and Reachable Sets

### 5.2.1 Airplane Dynamics

The model we use is given by the following set of five differential equations that describe a flight dynamics (see Figs. 5.1 and 5.2). This is a simplified flight dynamics. It is supposed that the aircraft is in the in-route flight. The influence of the wind is neglected. The angle of attack and the flight path angles are assumed to be small. The thrust and drag are supposed to be aligned. The initial value of the inertial heading is set equal to 90°. The fuel consumption is neglected during the maneuver and the air density is supposed to be constant. Consult Hull [6].

$$
\begin{cases}
\dfrac{dv}{dt} = \dfrac{T - D}{m} - g\gamma \\[2mm]
\dfrac{dh}{dt} = v\gamma \\[2mm]
\dfrac{d\psi}{dt} = \dfrac{L sin(\varphi)}{mv} \\[2mm]
\dfrac{dx}{dt} = v\, sin(\psi) \\[2mm]
\dfrac{dy}{dt} = v\, cos(\psi) \\[2mm]
L\, cos(\varphi) = mg,
\end{cases}
\tag{5.1}
$$

where
$T$—thrust,
$D$—drag,

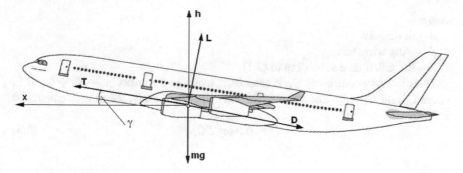

**Fig. 5.1** Aircraft side view

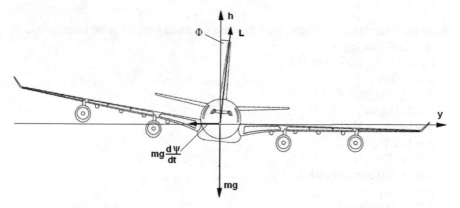

**Fig. 5.2** Aircraft front view

$L$—lift,
$\varphi$—aerodynamic bank angle,
$\gamma$—flight path angle,
$\psi$—inertial heading,
$x$, $y$—horizontal position ($x$-forward),
$h$—vertical position.

The definition and overview of differential inclusions can be found in Chap. 3. In this chapter, we use the flight controls to generate the corresponding differential inclusion. Then, the inclusion is solved using the differential inclusion solver (see Chap. 3, Sect. 3.6). This provides the attainable sets for the aircraft maneuvers.

Variables $T$, $\gamma$, and $\varphi$ are not exactly defined, only the restrictions for these variables are given. Thus, when $T$, $\gamma$, and $\varphi$ scan their limits, the right-hand side of (5.1) scans a set in the five-dimensional state space. In this way, we obtain a set $F(x, t)$ (the right-hand side) of the corresponding differential inclusion. The solution to this inclusion (the reachable set) shows the possible range of the flight state variables.

The drag $D$ is calculated as follows. First, the coefficient $C_L$ is calculated from the equation

$$L = 0.5 \, \rho v^2 S C_L, \tag{5.2}$$

where

    $\rho$—air density,

    $S$—gross wing area,

    $L$—lift, calculated according to (5.1).

Then, the drag coefficient $C_D$ is calculated from the formula $C_D = X_0 + C_{DT}C_r^2$, where $C_0$ and $C_{DT}$ are known from the aircraft data. Finally, the drag is calculated as follows:

$$D = 0.5\,\rho v^2 SC_P. \tag{5.3}$$

## 5.2.2 Attainable Sets

In the experiments shown here, the following parameter values have been assumed:

    $m = 200,000\,\mathrm{kg}$,

    $C_0 = 0.018,\ \ C_{DT} = 0.0342$,

    $S = 353\,\mathrm{m}^2$,

    $T = 1,200,000\,\mathrm{N}$,

    $\rho = 0.5\,\mathrm{kg/m}^2$,

    $\gamma \in [-18°, +18°]$,

    $\varphi \in [-21°, +21°]$,

    $v(0) = 200\,\mathrm{m/s}$,

    $h(0) = 0$ (relative height),

    $\psi = 90°$,

    $x(0) = 0\ \ y(0) = 0$.

In the following figures, the state variables are as follows:

    $x_1 = v,\ \ x_2 = h,\ \ x_3 = \psi,\ \ x_4 = x,\ \ x_5 = y$,

Figure 5.3 shows the *reachable set* (RS) for the flight trajectories, coordinates $y$ and $h$. These are the trajectory end points reached after 15 s of flight. Each dot represents a position that belongs to the boundary of the reachable set. Some points seem to be inside the RS, but this is not true. Note that what we see on a 2D image is only a projection of a five-dimensional cloud of points. Observe the small cluster of points (small dots) in the center of the figure. This set is the result of simple random shooting, where the control variables $(\gamma, \varphi)$ are generated as random ones, within the same limits. The cluster contains 10,000 trajectory end points, while the RS obtained with the DI Solver consists of only 5000 trajectories. We can appreciate how inefficient is the simple (primitive) random search.

In Fig. 5.4, there is a 3D image of the same RS, shown as a cloud of points, in two different view angles. On the right image, bold lines have been manually added to make the shape of the cloud much visible.

If we increase the limits for the control variables and the final flight time, then the RS becomes much deformed due to the non-linearity of the model. In Fig. 5.5, you can see the image similar to that of Fig. 5.3, with the final flight time equal to 30 s. In the static image, the 3D shape may not be very clear. The DI solver displays these images rotating around selected axes, so the 3D shape is better seen.

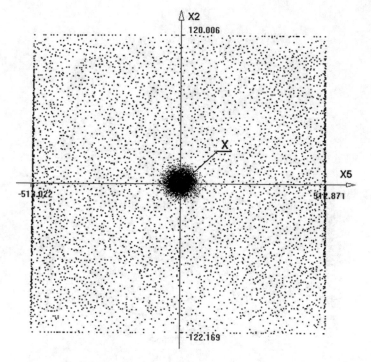

**Fig. 5.3** The reachable set for time = 15, coordinates y and h

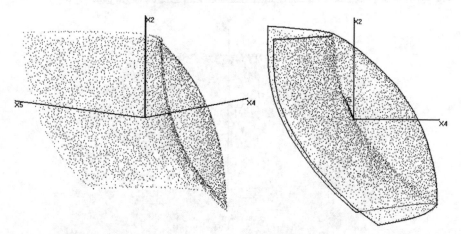

**Fig. 5.4** 3D images of the reachable set

The system trajectories are being stored in a file, together with the respective controls. Thus, selecting any point from the RS image we can see not only the model trajectory, but also the corresponding strategies (controls). Figure 5.6 shows a 3D view of the reachable set with model parameters as for Fig. 5.5.

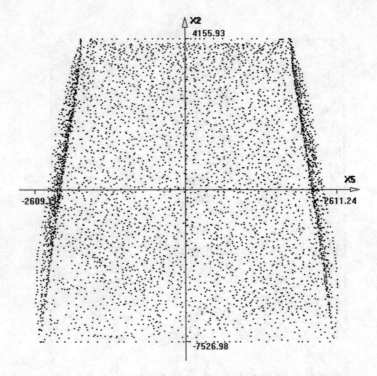

**Fig. 5.5** The reachable set after 30 s of flight increased limits for the flight path angle

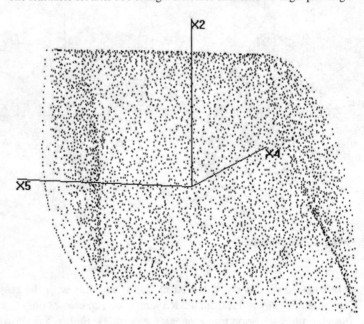

**Fig. 5.6** 3D view of the reachable set with model parameters as for Fig. 5.5

## 5.3 Conclusion

In the flight control, the knowledge about the reachable set may be crucial for safety problems. Parameter uncertainty analysis is a necessary task in robust control design. The DI solver, implemented on new parallel processing hardware, could be used in such practical applications.

The DI solver works satisfactory for the above and similar problems. For the models of dimensionality greater than two, some difficulties may arise with the results display. The solver produces a cloud of points that belong to the boundary of the reachable set. However, what we can see on the screen are only the two-dimensional projections of the cloud. A virtual reality technique and the animation may help in RS visualization.

## 5.4 Questions and Answers

**Question 5.1** What kind of flight dynamics model is used in this chapter?
a. Linear multivariable differential equations,
b: A set of non-linear ordinary differential equations,
c: The fluid dynamics partial differential equation for the air flow around the craft,
d: A system dynamics (SD) model given in form of SD block diagram scheme.

**Question 5.2**
Why do we use differential inclusions in this research?

**Question 5.3**
What do we assume about the aircraft movement and the main model variables?

**Question 5.4**
With respect to which control variables the reachable set is calculated in this chapter?

**Question 5.5**
Why the images of the reachable sets show areas like clouds of points, instead of contours that indicate the set boundary?

**Question 5.6**
To obtain the reachable set, are the fluctuations of control variables $\gamma$ and $\varphi$ generated randomly?

**Answers**

**Answer 5.1**
b: A set of non-linear ordinary differential equations.

**Answer 5.2**
The main topic of this chapter is the attainable sets for flight trajectories. The differential inclusion is used to calculate the attainable sets, using the differential inclusion solver.

**Answer 5.3**
It is supposed that the aircraft is on the in-route flight. The influence of the wind is neglected. The angle of attack and the flight path angles are assumed to be small. The thrust and drag are supposed to be aligned. The initial value of the inertial heading is set equal to 90°. The fuel consumption is neglected during the maneuver and the air density is supposed to be constant.

**Answer 5.4**
The control variables are aerodynamic bank angle and flight path angle.

**Answer 5.5**
What we see are the end points of the trajectories that are located at the boundary of the reachable sets. We do not see contours because what is displayed are projections of a five-dimensional cloud of the boundary points (like a five-dimensional "balloon surface") into a two-dimensional plane.

**Answer 5.6**
No, the controls are not generated as random variables. The generation of these functions is controlled by the *differential inclusion solver* in such a way that the trajectories scan only the boundary of the reachable set, and not its interior. Consult Sect. 3.6 of Chap. 3.

# References

1. Berkovitz LD (1964) A variational approach to differential games. In: Shapley LS, Tucjer AW (eds) Advances in game theory. Princeton University Press, Princeton, NJ
2. Dutton KE (1994) Optimal control theory determination of feasible return-to-launch-site aborts for the HL-20 personnel launch system vehicle. Report NASA technical paper 3449 Langley Research Center
3. Fahroo F, Ross IM (2001) Second look at approximating differential inclusions. J Guidance Control Dyn 24(1):131–134
4. Fudenberg D, Tirole J (1991) Game theory. MIT Press
5. Grigorieva SV, Ushakov VN (2000) Use finite family of multivalued maps for constructing stable absorption operator. Topol Methods Nonlinear Anal Juliusz Schauder Center 15(1)

6. Hull D (2007) Fundamentals of airplane flight mechanics. Springer, Berlin. ISBN: 10 3-540-46571-5
7. Issacs R (1999) Differential games. Dover Publications Inc., New York
8. Mordukhovich BS (2007) Optimal control of nonconvex differential inclusions. Differential equations, chaos and variational problems. Springer, pp 285–303. https://doi.org/10.1007/978-3-7643-8482-1-23
9. Petrosjan LA, Zenkevich NA (1996) Game theory. World Scientific Publishing Co., Inc, London
10. Raivio T (2000) Computational methods for dynamic optimization and pursuit-evasion games. Research reports, A90, Helsinki University of Technology, Systems Analysis Laboratory
11. Seywald H (2003) Desensitized optimal trajectories with control constraints. In: Conference paper: proceedings of the AAS/AIAA space flight mechanics meeting, Paper no. AAS 03-147, Ponce, Puerto Rico
12. Seywald H (1994) Trajectory optimization based on differential inclusions. J Guidance Control Dyn 17(3):480–487
13. Solan E, Vieille N (2001) Quitting games. Math Oper Res 26:265–285

# Chapter 6
# Modeling, Simulation, and Optimization

## 6.1 Introduction

This chapter is dedicated to optimal control. This topic is included in the book because the methods of optimal control theory are closely related to the differential inclusions, in particular, to the problem of finding the reachable sets. An application in marketing is presented.

Dynamic market optimization with respect to price, advertisement, and investment is considered. The model is non-linear. Its main parameters are the elasticities with respect to price, advertisement, and consumer income. Dynamic elements have been added to the static model. The parameters like seasonal index and consumer income are functions of time, and the whole market grows due to the investment. The tools of the optimal control theory are applied to calculate optimal policy for product price, advertisement, and investment, controlled simultaneously. The total revenue is maximized.

Here, we will discuss some practical aspects of the maximum principle of Pontryagin [19] and give an example of, perhaps not so typical, application in marketing.

Recall the statement of the basic optimal control problem. Consider a control system described by the following (vectorial) equations:

$$\frac{dx}{dt} = f(x(t), u(t), t), \tag{6.1}$$

where $x \in R^n$, $u \in C(x(t), t) \; \forall \, t \in [0, T]$, $T > 0$. Denote $x_0 = x(0)$.

Suppose that $f$ satisfies the Lipschitz condition and has continuous partial derivatives with respect to the state vector $x = (x_1, x_2, x_3, ..., x_n)$ and to the control vector $u = (u_1, u_2, u_3, ..., u_m)$. $C(x, t)$ is the set of control restrictions, $t$ represents the time, and $T$ is the final time of the control process. With such assumptions, for any function $u(t) \in C(x(t), t)$ there exists a function $x$ that satisfies (6.1), called *admissible trajectory*. The problem is to find a control function $u(t) \in C(x, t) \, \forall \, t \in [0, T]$ that minimizes the cost functional

© The Author(s), under exclusive license to Springer Nature Switzerland AG 2022
S. Raczynski, *Models for Research and Understanding*, Simulation Foundations,
Methods and Applications, https://doi.org/10.1007/978-3-031-11926-2_6

$$J = \psi(x(T)) + \int_0^T \phi(x(t), u(t), t)dt, \tag{6.2}$$

where $\phi$ is a continuous function with continuous derivatives with respect to $u$ and $x$. In the case considered in this chapter, the component $\psi$ is not needed ($\psi \equiv 0$). In a more general case, the derivative of $x$ may also appear as an argument of $F$. The control $u$ is restricted as follows:

$$u(t) \in C(x(t), t) \, \forall \, t \in [0, T]. \tag{6.3}$$

Here, we restrict the problem to the case without equality constraints. The conjugated vector $p = (p_1, p_2, ...., p_n)$ is introduced. By definition, it obeys the following equation:

$$\frac{dp_i}{dt} = -\sum_{j=0}^{n} \frac{\partial f_j}{\partial x_i} p_i - \frac{\partial \phi(x(t), t)}{\partial x_i}. \tag{6.4}$$

The *Hamiltonian function* for this problem is defined in the following form:

$$H \equiv p^T f(x, u, t) + \phi(x, u, t), \tag{6.5}$$

where $p^T$ is the transpose of $p$. In the classical definition of the maximum principle, the problem is to minimize the optimization criterion. According to Pontryagin's maximum principle [19], the optimal control which minimizes the cost functional $J$ over the time interval $[0, T]$ must maximize $H$ for almost all $t \in [0, T]$. Thus, the problem of minimizing the object function can be replaced by the problem of finding the maximum of $H$ with respect to $u$, with the restriction $u \in C(t)$, for almost all $t \in [0, T]$. The maximum principle gives the *necessary condition* of optimality.

The last problem of maximization of the Hamiltonian is easier than the original one. Let us give a heuristic, rather intuitive explanation. Let us discretize the problem in time, considering $k$ discrete time steps. Let the number of numerical integration steps over the interval $[0,T]$ is equal to $k = 1000$, and the control variable dimensionality $m = 3$. Then, the original problem is to minimize $J$, which is a function of $k \times m = 3000$ variables. Using the maximum principle, we decompose the original task into $k$ separate tasks of maximization a function (Hamiltonian) of only three variables. Thus, we repeat a simple optimization procedure $k$ times, but never solve any optimization problem with 3000 variables.

In some cases, the optimal control can be found analytically. If this is impossible, iterative algorithms may be used, as described later on. Let us see a simple example of analytic solution to an optimal control problem (Fig. 6.1).

**Fig. 6.1**  Landing on the moon

## 6.2   Landing on the Moon

Suppose that the landing module is suspended over the moon surface, $h = 500$, with no movement. It has two rocket engines which can provide a thrust $P$ accelerating downward or upward. The total thrust $P$ is limited: $-P_m \leq P \leq P_m$.

The problem is to land on the surface in minimal time. To avoid a crash, the final velocity must be equal to zero. The fuel consumption is also taken into account in the optimality criteria. Only vertical movement is considered.

During landing, the module position $h$ obeys the following equation:

$$\frac{d^2h}{dt^2} = \frac{P(t)}{M} - g. \tag{6.6}$$

Positive velocity is oriented upward. $M$ is the mass of the module (supposed to be constant) and $g$ is the moon gravity acceleration equal to $1.625 \, \text{m/sec}^2$. We introduce the state and control variables: $x_1 = h$, $x_2 = dh/dt$, and $u = u_1 = P$. Thus, the movement equations became as follows:

$$\begin{cases} \dfrac{dx_1}{dt} = f_1(t) = x_2(t) \\[3mm] \dfrac{dx_2}{dt} = f_2(t) = \dfrac{u(t)}{M} - g. \end{cases} \tag{6.7}$$

The landing final time is equal to $T = \int_0^T 1 dt$. Consequently, our object function is given by the following expression:

$$J = \int_0^T (1 + k|u(t)|)\, dt, \tag{6.8}$$

thus, $\phi(x, u, t) = 1 + k|u(t)|$. Consequently, we have

$$\frac{\partial f_1}{\partial x_1} = 0, \ \frac{\partial f_1}{\partial x_2} = 1, \ \frac{\partial f_2}{\partial x_1} = 0, \ \frac{\partial f_2}{\partial x_2} = 0, \ \frac{\partial f_1}{\partial u} = 0, \ \frac{\partial f_2}{\partial u} = 1/M. \tag{6.9}$$

According to 6.4, the equations for the conjugated vector are as follows:

$$\frac{dp_1}{dt} = 0 \text{ and } \frac{dp_2}{dt} = -p_1. \tag{6.10}$$

This means that $p_1$ is constant, and $p_2(t) == -p_1 t$.
Thus, the Hamiltonian is as follows:

$$H = p_1 x_2 - p_1 t \left( \frac{u(t)}{M} - g \right) + 1 + k|u(t)| \tag{6.11}$$

with $p_1$ and $k$ constant.

First, suppose that $k = 0$ (we don't care about the fuel consumption). Recall that we must maximize $H$ with respect to $u$ over the interval $[0,T]$. Observe that if $p_1$ is negative, then $u_1$ must be equal to $P_m$, otherwise it must be equal to $-P_m$ to maximize $H$. In our case, $p_2$ can change sign only once, so we get the "bang-bang" type of control with only one switch point or no switching at all. In fact, this is the only one, but important conclusion we get from the maximum principle. As explained below, for $k > 0$, we will have two switching points, and the control is equal to zero within some interval of time.

For any constant $u$, the model trajectory is a parabola on $(x_1, t)$ plane (Eq. 6.6). Let $k = 0$. If there were no switch, then we would have only one parabola and the module can only go upward or go downward and crash with great final velocity. So, we must have one control switch. This means that the whole trajectory consists of two parabolic segments, as shown in Fig. 6.2. Note also that $x_2(t)$ must be a continuous function, so $x_1(t)$ must have the continuous derivative.

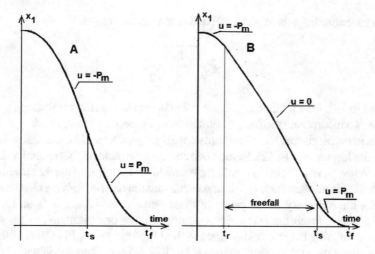

**Fig. 6.2** Landing trajectories

The solution for $x_1$ is as follows:

$$\begin{cases} x_1(t) = h - \left(\dfrac{P_m}{M} + g\right)\dfrac{t^2}{2} \text{ for } 0 \le t \le t_s, \\[4mm] x_1(t) = \dfrac{1}{2}\left(\dfrac{P_m}{M} - g\right)(t_f - t)^2 \text{ for } t_s \le t \le t_f, \end{cases} \tag{6.12}$$

where $t_s$ is the switch time and $t_f$ is the final time of the movement. As $x_1$ is continuously differentiable, we obtain the following equations:

$$\begin{cases} h - \left(\dfrac{P_m}{M} + g\right)\dfrac{t_s^2}{2} = \dfrac{1}{2}\left(\dfrac{P_m}{M} - g\right)(t_f - t_s)^2 & (a) \\[4mm] \left(\dfrac{P_m}{M} + g\right)t_s = \left(\dfrac{P_m}{M} - g\right)(t_f - t_s) & (b). \end{cases} \tag{6.13}$$

Equation (a) of (6.13) results from the continuity of $x_1$ at $t = t_s$, and (b) holds because $x_1(t)$ is continuously differentiable. From the above two equations, we can calculate $t_s$ and $t_f$. From (6.13 (b)), we obtain

$$t_f = \frac{2P_m}{P - Mg}t_s. \tag{6.14}$$

After substituting this to (6.13 (a)) and rearranging, we get

$$t_s = \sqrt{\frac{hM}{P + Mg}}. \tag{6.15}$$

From (6.14), we calculate $t_f$. Figure 6.2a shows the resulting trajectory.

Now, we incorporate the fuel consumption to the cost function, i.e., $k > 0$. Recall that we must maximize the Hamiltonian with respect to $u$. The control $u$ appears also in the last term of (6.12). Remember that $p_2$ intersects the horizontal axis (zero value). When $p_2$ is sufficiently small, the term $k|u|$ dominates, so the maximum of the Hamiltonian is reached for $u = 0$ because this maximizes the negative absolute value of $u$. So, the trajectory is composed of three parts, $u = -P_m$, $u = 0$, and $u = P_m$ (see Fig. 6.2b). From the continuity assumptions, we can calculate, in the similar way, the switching times $t_r$ and $t_s$, and the final time $t_f$. This will require somewhat complicated geometrical considerations, but this is rather geometric and algebraic problem.

## 6.3   Iterative Algorithm

In more complicated, multi-dimensional case, the (numerical) problem is that we do not have the initial conditions for vector $p$ and we cannot integrate (6.4) forward in time because of the lack of initial conditions for $p$. Instead, we have the final conditions for $p$. These conditions can be derived from the *transversality conditions*, consult Lee and Markus [13]. This leads to the known *two-point boundary problem*. Anyway, the control $u$ is still unknown, so we cannot integrate the state and conjugated equations. A possible solution consists in assuming an arbitrary admissible control, integrate the equations, and then looking for a way to improve the control. The process is repeated with the new control function. To be able to improve the control, we must know the direction in which the control should change. This direction is given by the gradient of the Hamiltonian (we try to maximize it).

The gradient of $H$ is given by the following expression:

$$grad\, H = p \cdot g, \text{ where } g = \left( \frac{\partial f_1}{\partial u_1}, \frac{\partial f_2}{\partial u_2}, ..., \frac{\partial f_n}{\partial u_n} \right) \tag{6.16}$$

for each time step.

In other words, $grad\, H$ is, at the same time, the search direction in the control space. To calculate $grad\, H$ for all $t \in [0, T]$, a trajectory $x(t)$ must be calculated and stored, starting with a given initial condition for $x$ and an arbitrary control. After this, the conjugated vector equations are calculated backward, starting with the final conditions for $p$.

According to the transversality conditions for the problem, (fixed time, free end point), the final condition for the conjugated vector is, in our case, $p_1(T) = 1$, $p_2(T) = 0$, ...., $p_n(T) = 0$. A possible iterative algorithm is as follows (see Polak [18]):

*1. Select an arbitrary control function u(t) (in fact, it should be an approximation of an optimal control, if we can get one), integrate Eq. (6.1) over [0,T], and store the state for each time step.*

*2. Starting with the final condition for p, calculate and store the trajectory of p(t) backward in time, using Eq. (6.4). Simultaneously, calculate and store gradH.*

*3. Once gradH has been defined, adjust control u in the search direction. Now, having this new control function, go to step 1.*

The stop condition for this algorithm can be to achieve $grad\,H$ small enough or to detect a lack of improvement of the object function.

This is a simplest possible version of the optimization algorithm. To accelerate the search, we can replace $u$ in step 3 by the new value which maximizes the Hamiltonian, instead of advancing a small step in the search direction. This, however, may provoke stability problems for non-linear models. Other common modification is to implement a steepest descent with conjugated gradients in the control space (Polak [18]). In our case, a simple steepest descent method was implemented.

## 6.4   Market Optimization

The market is a complex, non-linear, stochastic, socio-economic system. Such objects are known as *soft systems*. Looking at the annals of the huge literature in the field and comparing the models used, one can observe that the models almost always are completely different from each other. Many publications in the field of marketing, like those of Lilien and Kotler [14], offer a large number of models (linear, non-linear, deterministic, stochastic, static, dynamic, etc.). Any model should be related to the corresponding experimental frame (Zeigler [25]). To be realistic, the experimental frame must include the most important marketing parameters.

The model must be clear enough to be understandable for marketing and management staff, and it must be useful while carrying out simulation experiments. If we also want to find the optimal control of the market, then the simulations must be embedded in optimization algorithms and must be fast enough. Finally, another requirement is the ability to simulate the market dynamics. This means that the optimization must be dynamic, like in optimal control problems. In this chapter, a non-linear model with elasticities with respect to price, advertisement, and consumer income has been modified with dynamic inertia added to the market response. The resulting model is subject to multivariable dynamic optimization. Other important variables are included, like seasonal index, overall market growth, and investment, being functions of time.

The search for market models for simulation purpose dates from early 1960s, while significant publications began to appear in late 60s. King [10] gives a comparison of iconic, analog, and symbolic models. Recall that an *iconic model* represents reality on a smaller scale, *analogical model* shows reality in maps and diagrams, and a *symbolic model* uses mathematical expressions to portray reality. Montgomery and Urban [15] discuss the *descriptive*, *predictive*, and *normative models*, where by descriptive we mean models that consist largely of diagrams and maps or charts designed to describe a real-world system. See also Stanovich [23]. Predictive model is used in predictive analysis to create a statistical model of future behavior, and normative model evaluates alternative solutions in order to answer the question "What is going on?" and to suggest what ought to be done or how things should work.

We should distinguish between macromarketing and micromarketing models (Lilien and Kotler [14]). Macromarketing addresses big and important issues at the nexus of marketing and society, while micromarketing refers to marketing strategies, which are customized to either local markets, to different market segments, or to the individual customer, see Shapiro et al. [20]. Here, we deal with the symbolic and micromarketing model that can be implemented in computer simulations and then used in optimal control algorithm.

Optimal control in marketing is not new. There are many publications in the field like, for example, Bertsimas and Lo [1]. In that article, we can find the application of the Bellman's dynamic programming approach to the problem of the price impact on the dynamic trading on the stock exchange. A tutorial and survey on the relevant technical literature on models of economic growth can be found in Burmeister and Dobell [2]. Feichtinger et al. [5] consider a similar problem for a general market and optimal advertising policy. In that article, a detailed formulation of an implementation of Pontryagin's maximum principle is shown. Yuanguo [24] uses Bellman's principle of optimality and derives the principle of optimality for fuzzy optimal control. This is applied to a portfolio selection model.

The general approach to optimization of economic systems including households' consumption, labor supply, production, and government policies can be found in the book of Dixit [4]. As the author explains, *"the methodology is based on the use of verbal and geometric arguments, but with the eye toward the mathematical sharpening and generalization. The concepts of optimizing with respect to such variables as prices, consumer income and quantities of goods can make this methodology useful in marketing problems."*

Chow [3] in his book describes the application of the Lagrange method to the optimization of economic systems in general. Instead of using dynamic programming, the author chooses the *method of Lagrange multipliers* in the optimization task. A number of topics in economics, including economic growth, macroeconomics, microeconomics, finance, and dynamic games, are treated. The book also teaches by examples, starting with simple problems. Then, it moves to general propositions.

The Lagrange method is a powerful optimization tool, but needs some additional conditions and modifications to be applied in practice. Namely, optimization of models with constraints should be treated rather with the tools of the optimal control theory, like the maximum principle used in this chapter.

Konno and Yamazaki [11] present a large-scale optimization problem of a stock market with more than 1,000 stocks and show that the problem can be solved using the *absolute deviation risk* (called L1 risk function). In fact, stock market has its own properties that need methods oriented to that particular problem.

Other approach to the stock market optimization can be found in Speranza [21]. The paper describes an application of an optimization algorithm to the Milan stock market, taking into account portfolio with transaction costs, minimum transaction units, and limits on minimum holdings. The author points out that the presence of integer variables dramatically increases the computational complexity.

Karatzas et al. [8] consider a general consumption/investment problem for an agent whose actions cannot affect the market prices, and who strives to maximize total expected discounted utility of the consumption as well as terminal wealth. They decompose the problem by considering separately maximizing utility of consumption only, and of maximizing utility of terminal wealth, and then appropriately composing them. Such decomposition may work in some special cases. However, in a general case, it may not work. In the present article, we do not assume any possibility of decomposing the optimization problem to any partial sub-problems.

In another paper of Karatzas [9], we can find a unified approach, based on stochastic analysis, to the problems of pricing, consumption/investment, and equilibrium in a financial market with asset prices modeled by continuous semi-martingales, and a similar problem decomposition. The Hamilton–Jacobi–Bellman equation of dynamic programming associated with this problem is reduced to the study of two linear equations. The results of this analysis lead to an explicit computation of the portfolio that maximizes capital growth rate from investment, and to a precise expression for the maximal growth rate.

Korn and Korn [12] in their book offer a collection of graduate studies in mathematics, including the mean-variance approach in a continuous-time market model, pricing of exotic options and numerical algorithms. The marketing models in the book are shown from the mathematical point of view and reflect the state of art in the field (as for year 2001). Gomes Salema et al. [7] contemplate generic reverse logistics and distribution network, where capacity limits, multi-product management, and uncertainty on product demands and returns are considered. A mixed integer formulation is developed, using standard B&B (Business-to-business) techniques. The model is applied to an illustrative case. To learn more on the B-to-B strategies, consult, for example, Morris et al. [16]. There are many other publications on optimal market control, most of them applied to the stock market or to markets of specific goods. It seems that the dominant tool is the dynamic programming and Lagrange method.

The market model itself is not the main topic of the present chapter (except, perhaps some dynamics added). Let us start with the model taken from the book of Lilien and Kotler [14], mentioned earlier. We discuss an application of maximum principle to the problem of dynamic optimization of a non-linear revenue model with respect to price, advertising, and investment. The elasticity-based model has been modified to reflect the market dynamics, including the inertia with respect to model response. This is a non-linear model of the demand and revenue. The descriptive variables are price, advertising, seasonal index, market overall growth, consumer income, production or acquisition cost, and the market elasticities with respect to the price, advertising, and consumer income. The demand model is as follows:

$$q(p, a, y, v, g, t) = q_0 v(t) s(t) \left( \frac{p(t)}{p(0)} \right)^{e_p} \left( \frac{a(t)}{a(0)v(t)s(t)} \right)^{e_a} \left( \frac{y(t)}{y(0)} \right)^{e_y}, \quad (6.17)$$

where $q$—demand, $q_0$—initial or reference demand,

$t$—the time,

$s(t)$—overall market size,

$p(t)$—price,

$a(t)$—advertising per time unit,

$y(t)$—consumer income,

$v(t)$—seasonal index,

$e_p, e_a, e_y$—market elasticity with respect to price, advertising, and consumer income, respectively.

In the model of Lilien and Kotler, the market size $s(t)$ is supposed to be equal to $(1 + g)^t$, where $g$ is the market growth rate. We use the variable $s(t)$ instead, to be able to link the market growth to the investment that the company may do in order to expand the market (new installations, infrastructure, etc.).

Figure 6.3 shows the shape of the profit (utility) $(p - c)q$ as a function of the price and advertising, when the time and other variables are fixed ($c$ is the unit cost of product).

Normally, the price elasticity is negative (price increase means less demand), and the elasticity for advertising and consumer income are positive. Note that the term $v(t)s(t)$ multiplies the demand and appears also in the denominator of the advertising impact term. This means that if the market and the seasonal index grow, then we must

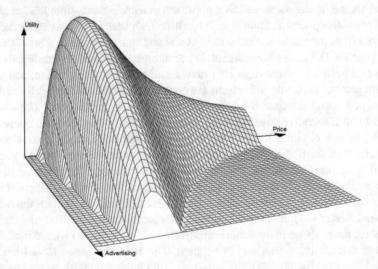

**Fig. 6.3** The revenue as function of the price and advertising. Static case

spend more on advertising to achieve the same effect. We do not consider any storage or warehouse mechanism, so it is supposed that the sales are equal to the demand.

As for the competition, the model includes it in some indirect way. Observe that the market elasticities with respect to price and consumer income are influenced by the competition. If, for example, a company provides water to a village and has no competitors, then the elasticity $e_p$ will approach zero because the consumers must consume certain amount of water, anyway. However, if there is a competitor in the region, then this elasticity must be negative because the price will influence the demand of water provided by the company. Other way to introduce competitors to the model would be to simulate the performance of two or more competing companies. However, if we intend to optimize the market for one company, then we must simulate the optimal strategy of the competitors as well. This leads to a differential game and not just a single optimization problem.

There are many approaches to competition modeling. Perhaps one of the most representatives is the spatial competition model and the *hedonic or Lancaster model*, see Page [17].

In the spatial model, there is an array of *spatial attributes* A that characterizes the product, like the content of sugar of the vitamin C and price. The consumers have their own attribute vector of preferences B. The consumer payoff depends on the (Euclidean) distance $D(A, B)$ between the spacial attributes and preferences, expressed as $K - D(A, B)$, K being a constant. So, the customers try to minimize their distance from the preferred product and maximize the payoff. In the Hedonic model, the consumers are characterized by the weights they associate with the product attributes, and the payoff is the scalar product of the attributes and weights. We do not discuss spatial and hedonic models here because these are static models, and the present book treats about dynamic systems.

In the model used here, the revenue can be calculated as follows:

$$U = (I - b)e - a, \quad V = e - a, \quad \text{where } e = (p - c)q. \tag{6.18}$$

Variable $U$ will be called total revenue and $V$ just revenue. Here, $c$ is the unit cost of the product (production, acquisition cost), $a$ is the advertising, $b$ is the investment factor, and $q$ is given by formula (6.17). The investment $b$ is supposed to be a part of the utility $e$, $0 \le b \le 1$. The factor $b$ defines the market growth rate, proportional to the product $be$. This will be one of our control variables. In our equations, the market size is relative with initial value equal to 1. This means that if the company does not invest in the market, then the market size has no influence on the model trajectory (remains constant). The final total revenue is the sum (integral) of $U$ over the time interval under consideration. Figure 6.3 shows the shape of the profit (utility) $e$ as a function of the price and advertising, while the time and other variables are fixed. It can be seen that the utility function has a maximum that determines the optimal price and advertising at the moment. This is a static case.

Now, consider a fixed time interval $[0, T]$. To calculate the total revenue, we must integrate $U$ over $[0, T]$. We should also take into account the market inertia with respect to the price and advertisement. In our model, we added the first-order inertia with time constants $T_p$ and $T_a$ , for the price and advertisement variables, respectively. So, the model becomes of the fourth order, described by the following equations:

$$
\begin{cases}
\dfrac{dr}{dt} = (1 - b(t))v & \text{total revenue} \\[2mm]
\text{where } v = (p(t) - c(t))q(z, w, y, v, s, t) - a(t) \\[2mm]
\dfrac{dw}{dt} = (a(t) - w(t))/T_a & \text{advertisement inertia} \\[2mm]
\dfrac{dz}{dt} = (p(t) - z(t))/T_p & \text{price impact inertia} \\[2mm]
\dfrac{ds}{dt} = kb(t)(p(t) - c(t))q(z, w, y, s, t), & \text{market size}
\end{cases}
$$

$$(6.19)$$

where $r$ is the total revenue, $w$ is the advertisement inertia called also *consumer goodwill* related to advertising, and $z$ is the price impact with inertia. The coefficient $k$ tells how fast the market grows due to the investment, its dimension is 1/(currency unit). For example, $k = 0.001$ means that one invested currency unit makes the market grow by 0.001(relative) in one time unit. The investment in the market growth cannot be negative, so the control $b$ in (6.19) is set equal to zero when $(p - c)q$ non-positive. It is supposed that $0 \le b \le 1$.

Note that in (6.19) the demand depends on $z$ and $w$ instead of the price $p$ and advertising a taken at the current time instant. The object functional we want to maximize is the final accumulated revenue $r(T)$, and the optimization is carried out with respect to the price $p$, advertising $a$ (through the corresponding controls, defined later), and the investment factor $b$. This way we obtain a system of four non-linear differential equations with three control variables. The variables $r$ and $s$ are closely connected to each other and also depend on w and $z$. The model is rather simple, but its optimization is not a trivial task.

In the optimization algorithm, the control variables are as follows:

$u_1$—tells which part of initial (reference) advertisement is used as the actual advertisement, $a = u_1 a_0$;

$u_2$—tells which part of initial (reference) price is applied;

$u_3$—tells which part of the revenue $V$ (6.18) is used as the investment (investment factor b).

The advertisement $a(t)$ does not depend explicitly on the investment factor $u_3$. However, the model includes a number of restrictions that result from the market logic. The price, advertisement, and investment cannot be negative. Moreover, the advertisement cannot be greater than $U - (1 - b)e$ . This makes the advertisement depending on $U$ and on the investment factor $b$, through the corresponding restriction.

## 6.5 Computer Implementation: Simulation and Optimization

The program developed for this task permits a big variety of simulation and optimization scenarios. Our optimization process is iterative, and the model is non-linear of order four with up to three controls, so the whole process is rather slow, with several minutes of computing time needed. As we need rather qualitative results, the integration of the model trajectories has been done by a fast, but not very exact, Euler's method, with a reasonable time step (1000 steps in the simulation time interval). There was little difference observed between the results obtained by this method and the Runge–Kutta integration scheme.

We assume the same fixed time horizon equal to 365 (days) in all simulations. In addition to the model Equations (6.19), some simple arithmetic operations have been added, to avoid negative revenue growth. In other words, we do not consider the possibility of negative revenue at any moment, and all the advertisement and investment must come from the instant sales income. Such restrictions, as well as other non-negativity restrictions imposed on the controls, complicate the optimization process. Recall that one of the assumptions of Pontryagin's maximum principle is that the right-hand sides of Equations (6.19) are continuously differentiable. In our case, this is not exactly true. So, all the results should be treated as approximations of the optimal solution (anyway, the algorithm is iterative), and some of the resulting curves show certain irregularities (are not "nice and smooth").

The initial conditions for the state vector have been fixed to (0, 0, 0, 1) for total revenue $r$, advertisement goodwill $w$, price impact with inertia $z$, and the relative market size $s$, respectively. Let us show results of four experiments, where the market is optimized with respect to price, advertisement policy, price and advertisement, and finally for all three controls simultaneously. To see the market strategy in changing environment, the seasonal index is set equal to one, except of two intervals of time: (100, 128) and (160, 167) days. In the first interval, the index jumps to 3, and in the second it falls down to 0.5. The relative consumer income also changes; it is equal to one everywhere, except of the interval (200, 214), where its value is equal to 2.5. It is supposed that the company (the decision-makers) is aware of those changes before they occur. This is important assumption because the optimization algorithm has a "predictive" ability, namely, the corresponding control changes anticipate the changes of these two indices. Model fixed parameters are as follows:

Initial (reference) sales = 70,000 items,
Production cost = 0.8 CU (currency units),
Initial (reference) price = 1.0 CU,
Initial (reference) advertisement = 10,000 CU,
Market elasticity for price = −2,
Market elasticity for advertisement = 0.5,
Market elasticity for consumer income = 0.3,
Time constant for advertisement inertia (goodwill) = 14 days,
Time constant for price impact inertia = 2 days,

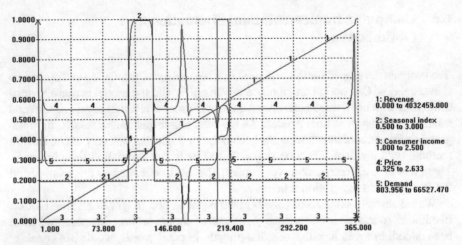

**Fig. 6.4** Price control only

Initial market size, relative = 1,

Final simulation time = 365 days,

Market growth factor = 0.00000005—this means that one invested CU makes the market relative size grow by 0.00000005 per time unit.

The set $C(t)$ of (6.3) is defined by the set of the restrictions imposed on the model. In particular, all model variables that, logically, cannot be negative are restricted to the non-negative values. The investment cannot be positive if the revenue $V$ is equal to zero. Also, the advertisement cannot be greater than $U - (1 - b)e$ (6.18).

This is an optimization with fixed time horizon. In other words, the company starts to sell goods and disappears after a given time interval. The final simulation time is equal to 365 days. The behavior of the control variables for the last few days appears to be somewhat strange. Obviously, the optimal control for investment and advertising near the end of the interval falls down; there is no reason to waste money if the activities terminate. On the other hand, the algorithm makes the price grow in the last days of the activity. The sales do not fall down immediately because of the market inertia with respect to the price, so we can get a little bit better final outcome. All curves are normalized to the interval [0,1]. The real ranges are indicated at the legend right to the plot.

Figure 6.4 shows the result of market optimization with respect to the price of the product. Note that when the seasonal index and the demand grow, the recommended policy is to lower the price. Anyway, the revenue (curve 1) grows faster in that period. When the seasonal index becomes low, the strategy is first to increase the price and then reduce it rapidly. As for the period of increased consumer income, the optimal policy is to lower the price to make the demand even higher. All these changes should be done with anticipation because of the market inertia.

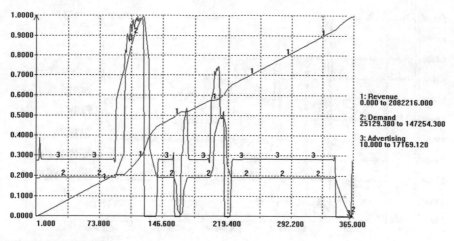

**Fig. 6.5** Advertisement control only

The result of advertisement policy optimization is shown in Fig. 6.5. In the period of augmented seasonal index, it is recommended to increase the advertising. When the consumer income grows, the advertising should grow as well.

If we optimize both the price and advertisement simultaneously, the curves are similar (Fig. 6.6). The total revenue at the end of the period is, in this case, greater than that of price and advertisement optimized separately. Figure 6.8 shows the comparison between the above optimization modes (revenue value).

The company may invest in the market (new marketing places, infrastructure). The income from sales can be invested or "consumed" immediately. If we invest, the market grows as well as the future income. Logically, the optimal policy should be to

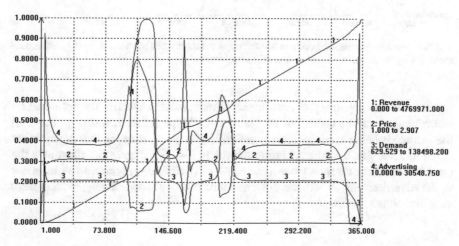

**Fig. 6.6** Price and advertisement control

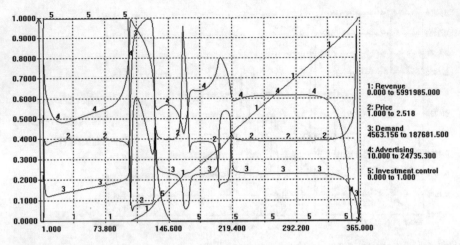

**Fig. 6.7**  Price, advertisement, and investment control

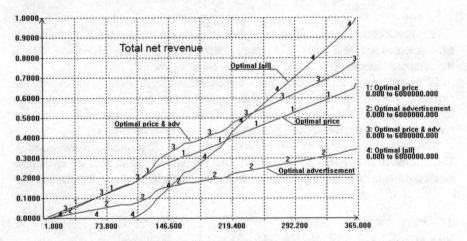

**Fig. 6.8**  Revenue curves. Comparison between optimization with respect to price only, advertisement, both price and advertisement, and all control variables

invest at the very beginning, in order to obtain more income later on. The optimization algorithm generates a control of the "bang-bang" type for the investment. First, all the income is being spent on investment and advertising, then the investment is set equal to zero. See the corresponding curves in Fig. 6.7. Figure 6.8 illustrates the comparison of the revenue growth for cases of optimization with respect to price; to the advertisement; to both the price and advertisement; and, finally, to all control variables simultaneously.

## 6.6 Conclusion

Control theory provides useful tools for dynamic optimization of models in many fields of research. Once we have the model equations, we can try to apply an optimization algorithm to it. The problem is that, in the real world, not everything obeys or can be modeled with differential equations. In marketing, optimization methods work to some extent, due to the intense research that has been done in past decades. This research resulted in a variety of models that can be used in dynamic simulation and optimization.

Though the presented application of optimal control theory does not use differential inclusions, it has been included in this book because we use some mechanism of the Maximum Principle in reachable set calculations. Anyway, each trajectory that scans the boundary of the reachable set is optimal in some sense. Recall that if a point in the model state space is accessible from the initial point by a trajectory of a dynamical system (with appropriate regularity assumptions), then it is also accessible in optimal (minimal) time, and the trajectory must entirely lie on the boundary of the reachable set.

The model itself is not the main topic of this chapter. However, its modification (price and advertising inertia) represents perhaps a relevant contribution. The main topic is the application of the methodology of the control theory (the maximum principle) to the multivariable dynamic market optimization. Our point is that, due to the restrictions imposed on control variables, Pontryagin's maximum principle is the most adequate tool. The algorithm and the experimental version of the software may form a core of a more practical and commercial marketing optimization package.

Recall that the optimization task in this chapter has been defined for a fixed time horizon. If we do not fix the time interval, supposing that the company will operate forever or for a large period of time, the results can be quite different. In such case, we should redefine the optimization criteria because the total revenue at the end of the period will be less adequate. Once this optimization goal is defined, a similar optimization tool can be applied. The further applications should include the multiple product case, stochastic case, and the marketing (differential) games with competition between multiple companies.

## 6.7 Questions and Answers

**Question 6.1** What is the *functional*? What criterion of optimization is used in this chapter?

**Question 6.2** What is the advantage of the Maximum Principle of Pontryagin?

**Question 6.3** What optimization criterion and conditions are used in the example of the optimal landing on the Moon?

**Question 6.4** The solution to our optimal landing problem is derived analytically or by an iterative algorithm?

**Question 6.5** How the iterative algorithm, used in market optimization, works?

**Question 6.6** What kind of demand model is used in this chapter (select):
1. A linear multivariable model.
2. A non-linear model based on market elasticities.
3. A multivariable regression model using market elasticities.
4. A partial differential equation model.

**Question 6.7** The demand model of Lilien and Kotler [14], Eq. (6.17), is a dynamic or static market model?

**Question 6.8** What represent the elasticities in market demand model?

**Question 6.9** What modification of the Lilien–Kotler [14] Eq. (6.17) model has been done in this chapter?

**Answers**

**Answer 6.1** A *functional* is a mapping from a space S into the field of real or complex numbers. In this chapter, S is the space of integrable functions of time. An example: the integral of a given function of time over a time interval is a functional.

In this chapter, we maximize a functional that represents the total utility of a market.

**Answer 6.2** In this method, the original problem of functional minimization (or maximization) is reduced to the problem of maximizing the *Hamilton function (Hamiltonian)* with respect to the control variable, subject to the control restrictions. Note that the Hamiltonian is a *function*, and not a *functional*.

**Answer 6.3** We minimize the time of landing, requiring that the final velocity of the lunar module is equal to zero. The total fuel consumption is also taken into account.

**Answer 6.4** This is an analytical solution.

**Answer 6.5** The algorithm calculates the gradient of the Hamiltonian with respect to the control function. Then, it improves the control function by maximizing the Hamiltonian in each time step. This procedure is repeated until satisfactory quasi-optimal control is determined.

**Answer 6.6** 2. A non-linear model based on market elasticities

**Answer 6.7** Static,

**Answer 6.8** We use the elasticity parameters $e_p, e_a, e_y$ that determine how the demand changes with respect to price, advertising, and consumer income, respectively.

**Answer 6.9** We added the first-order inertia with the price and advertisement variables. This makes the model more realistic, with an inertial delay in the market response.

# References

1. Bertsimas D, Lo AW (1998) Optimal control of execution costs. J Finan Markets 1:1–50
2. Burmeister E, Dobell AR (1972) Guidance and optimal control of free-market economics: a new interpretation. SMC 2(1):9–15
3. Chow GC (1997) Dynamic economics: optimization by the lagrange method. Oxford University Press, Oxford, p 9780195101928
4. Dixit AK (1990) Optimization in economic theory. Oxford University Press, Oxford
5. Feichtinger G, Hartl RF, Sethi SP (1994) Dynamic optimal control models in advertising: recent developments. Manage Sci INFORMS 40(2):195–226
6. Forrester JW (1961) Industrial dynamics. Pegasus Communications, Waltham, MA
7. Gomes Salema MI, Barbosa-Povoa AP, Novais AQ (2007) An optimization model for the design of a capacitated multi-product reverse logistics network with uncertainty. Eur J Oper Res 179(3):1063–1077
8. Karatzas I (1989) Optimization problems in the theory of continuous trading. SIAM J Control Optim 27(6):1221–1259
9. Karatzas I, Lehoczky JP, Novais AQ, Shreve SE (1987) Optimal portfolio and consumption decisions for a "small investor" on a finite horizon. SIAM J Control Optim Soc Ind Appl Math 25(6):1557–1586. http://dx.doi.org/10.1137/0325086, ISBN/ISSN ISSN 0363-0129
10. King WR (1967) Quantitative analysis for marketing management. McGraw-Hill, New York
11. Konno H, Yamazaki H (1991) Mean-absolute deviation portfolio optimization model and its applications to Tokyo stock market. Manag Sci 37(5):519–531
12. Korn R, Korn E (2001) Option pricing and portfolio optimization modern methods of financial mathematics, vol 31. American Mathematical Society, ISBN: 10: 0-8218-2123-7
13. Lee EB, Markus L (1967) Foundations of optimal control theory. Wiley. ISBN: 978-0898748079
14. Lilien GL, Kotler P (1972) Marketing decision making: a model-building approach. Harper & Row, New York, p 0060440767
15. Montgomery DB, Urban GL (1969) Management science in marketing. Prentice Hall, Englewood Cliffs, NJ
16. Morris MH, Pitt LF, Honeyutt ED (2001) Business-to-business marketing: a strategic approach. Sage, London. 0-8039-5964-8
17. Page SE (2018) The model thinker. Hachette Book Group, New York. 978-0-465-009462-2
18. Polak E (1971) Computational methods in optimization. Academic Press, New York, p 0125593503
19. Pontryagin LS, Boltyanskii VG, Gamkrelidze RV, Mishchenko EF (1962) The mathematical theory of optimal processes. Interscience. ISBN: 2-88124-077-1
20. Shapiro SJ, Tadajewski M, Shultz CJ (2009) Interpreting macromarketing: the construction of a major macromarketing research collection. Research collection. J Macromarket 29:325–334. https://doi.org/10.1177/0276146709338706
21. Speranza MG (1996) A heuristic algorithm for a portfolio optimization model applied to the Milan stock market. Comput Oper Res 23(5):433–441
22. Sriyudthsak K, Uno H, Gunawan R, Shiraishi F (2015) Using dynamic sensitivities to characterize metabolic reaction systems. Math Biosci 269:153–163
23. Stanovich KE (1999) Discrepancies between normative and descriptive models of decision making and the understanding/acceptance principle. Cogn Psychol 38:349–385. URL: http://www.idealibrary.com
24. Yuanguo Z (2009) A fuzzy optimal control model. J Uncertain Syst 3(4):270–279. URL: www.jus.org.uk
25. Zeigler BP (1976) Theory of modeling and simulation. Wiley-Interscience, New York

# Chapter 7
# Discrete Event Models

## 7.1 Introduction

Here, we mention some basic concepts of *Discrete event modeling and simulation*, necessary to well understand the remarks included in the other chapters. A greater application of discrete event and agent-oriented model is discussed in Chap. 8 (simulation of growing organizations and their interactions).

The discrete event models contain specifications of possible events that may occur in the modeled system. The order of event execution is not pre-established, and is defined at the runtime by the corresponding software, and not by the user. So, each one of the simulation tools like GPSS or Arena package is, in fact, certain version of discrete event model, ready to use for a class of similar modeling tasks, like queuing models and manufacturing. In discrete event modeling and simulation, the model and its software implementation can hardly be separated. Defining a model, we should define the possible event execution without fixing the time instants of execution. Almost all known discrete event simulation packages include, in fact, such model specification.

Creation of models and computer simulation are closely related, and sometime inseparable tasks. From this point of view, each one of simulation packages like GPSS, Arena, or ProModel represents its own modeling methodology. In discrete event simulation, the very elemental concept is that of object-oriented modeling (OOM). Recall that in OOM we create objects that consist of *data* and *methods*. The data describe the object private data structures, visible or not from other objects. The methods are computational procedures that process object data and/or execute object events. One of the first discrete event simulation packages of early 1960s was the CSL (control and simulation language) based on Fortran that used the concept of the *class of objects*. Then, an excellent and quite complete object-oriented language Simula67 appeared in 1967. This language, based on Algol68, ran on computers with 64K operational memory, 100 KHz CPU clock, but, from the point of view of programming methodology, it was perhaps never superated by the contemporary object-oriented languages.

© The Author(s), under exclusive license to Springer Nature Switzerland AG 2022          171
S. Raczynski, *Models for Research and Understanding*, Simulation Foundations,
Methods and Applications, https://doi.org/10.1007/978-3-031-11926-2_7

**Fig. 7.1** GPSS transactions and facilities

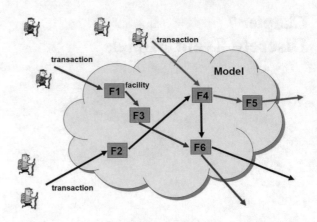

The idea of object creation and model event execution dates from 1960s and was well implemented in the GPSS package. The GPSS objects, named *transactions*, are created at the runtime. They pass through the model events (GPSS *resources*), interact with each other, and disappear. This concept repeats in nearly all discrete event simulation tools.

Transactions are moving objects that are generated during program run, pass through a number of *facilities*, and disappear. The GPSS facilities define possible events that occur in the "life" of a transaction, see Fig. 7.1. This concept is still used in other, more recent discrete event simulation software.

In GPSS, like in many others modeling tools, the events almost always occur is some random time instants, depending on the rules of interaction. We will not give here any overview or specification of the probability theory and statistics because this is rather the topic of textbooks on mathematics. Let only recall that the most important issue in the mass-service simulation is the correct definition of the distribution and parameters of the inter-arrival time intervals and service time. For example, if the clients arrive at a service facility according to the Poisson process (independently from each other), then the inter-arrival time interval should have the *exponential* distribution, and not the Poisson distribution.

In mass-service modeling, an important part is the modeling of waiting lines. Recall that the formation of queues has been a topic of research works for many years. The *queuing theory* was developed, based on the theory of probability. We will not discuss here the results of the queuing theory. The reason is that this book treats on the models of dynamic systems. The core of queuing theory offers a series of formulae for average queue length and other queue characteristics. In these formulae, the time variable does not appear.

## 7.2 The Event Queue

A discrete simulation language is as good as the algorithm that manages the *event queue*. This queue should not be confused with a queue we want to simulate, for example, a queue of clients in a mass-service system or a buffer in a manufacturing system where the parts wait to be processed. The event queue contains a set of *event messages*, each of them telling which model event to execute and specifying the time instant when the execution will occur. The advantage of discrete simulation is that the model time jumps to the next (in time) event to execute, instead of advancing continuously (in small time steps). This means that the system (the program, which controls the discrete event simulation) must know which the next event to execute is. There are two ways to achieve this. First, we can maintain the event queue always sorted due to the execution time, and second, to add new event messages at the end of the queue, and then look for the nearest event to execute. Both options involve the problem of sorting or scanning. This process is simple and fast if there are few events in the queue. However, if the model is not trivial (has more than, say, two queues and servers), the event queue can grow to hundreds or thousands of events, and the event handling strategy becomes crucial for the whole system performance.

Observe that the event queue in the simulation process constantly changes. Any event, while being executed, can generate one or more new event messages or cancel some of the existing ones (see Fig. 7.2). Moreover, there are events that cannot be scheduled through the event queue mechanism, being executed due to the changes of the model state and not of the model time. Such events are called *state events* and must be handled separately. There are three basic strategies in discrete simulation: *activity scanning* (AS), *event scheduling* (ES), and *process interaction* (PI). In this chapter, we treat *activity* and *event* as synonyms. More advanced strategies are being

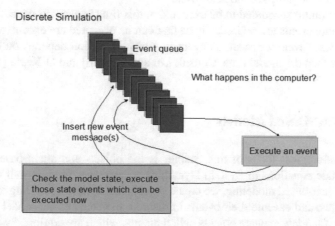

**Fig. 7.2** The event queue

developed and can be found in publications on the DEVS (discrete event specification formalism). See Chow [9] and Zeigler [67].

*Activity scanning* (AS) was the first discrete simulation strategy, developed in the 1950s. One of the first implementations was the language CSL. According to this strategy, the model time is set equal to the time instant of the nearest (*time*) event. Then, all model activities (events) are scanned. Those that can be executed are executed, and the others remaining inactive. Next, the time jumps to the next possible event, and the whole process is repeated. This clock loop stops if no possible events remain in the model. An event, while being executed, can schedule itself or other events to be executed in the future, so the sequence of executions can be long, even if the source program is relatively short.

The *Event Scheduling* (ES) strategy is somewhat more effective. In the computer memory, the *event queue* is created. Every component (*event message*) of this queue stores the time the event will be executed, and the event identifier. So, the only problem is to maintain the event queue sorted according to execution time. If we do this, then we simply take the first event and execute it, without scanning all possible events. This event queue management is transparent (invisible for the user) and works automatically. The user can schedule events, but he/she cannot redefine the model time or manipulate directly the event queue. The most effective event management algorithms are those using binary tree techniques to sort the event queue.

The *Process Interaction* (PI) strategy is more advanced. The model is defined in terms of processes that can run concurrently. The rules of interaction between processes are defined, while the specification of a process includes the necessary event scheduling. PI can be implemented in any object-oriented programming language, and became the main feature of Simula, Modsim, BLUESSS, and other languages.

The *three-phase strategy* in discrete simulation is a combination of these three strategies. The phases are as follows:

1. Model time jumps to the next event.
2. The event(s) scheduled to be executed at this time instant are executed.
3. All state events are revised. Those that can be executed are executed.

By the state event we mean an event in which execution depends on the model state rather than the model time. Consult Lin and Lee [36] and O'Keefe [46].

## 7.3  Agent-Based Models

The methodological focus of this chapter is the object- and agent-based simulation. No state equations or System Dynamics schemes are used. Recall that in the discrete object-based modeling, we create objects that behave according to the user-defined rules, and execute their events in discrete moments of the model time. The *agent-based models* manage objects called agents, which are equipped with certain "intelligence." They can take decisions, optimize their actions, and interact with each other and with the environment. Agent-based models (ABMs) are a type of microscale models of agents that simulate the simultaneous operations and interactions of multiple agents in an attempt to recreate and predict the global, complex phenomena.

The individuals in ABM models may be of different types. Although the rules of behavior are the same for individuals of the same type, the behavior is not identical for all of them. This modeling method has many applications, mainly in ecology, biology, and social sciences. The key notion is that simple behavioral rules (micro-model) generate complex (macro) behavior. An important central tenet is that the whole is greater than the sum of the parts. Individual agents are typically characterized as rational. They are presumed to be acting in what they perceive as their own interests, such as reproduction, economic benefit, or social status, using heuristics or simple decision-making rules (Railsback et al. [52], Bandini et al. [3]). However, the agents may also be irrational, in particular, when they act in groups. The collective behavior is frequently irrational, what can be observed clearly in voting activities.

Note the main difference between object-oriented and simulation package. The latter, in addition to object creation, provides (or should provide) a "clock" mechanism that automatically manages the model time and the event execution. The ABM modeling is supported by many programming and simulation tools. Let us mention only some of the most popular tools: SWARM developed in 1994 by the Santa Fe Institute (Swarm Development Group, [59]), Ascape program developed in 2001 (Parker [48]), Breve-2.7.2 (Klein [33]), Recursive Porous Agent Simulation Toolkit released in 2003 (Michael et al. [45]), Cormas developed in 2004 by VisualWorks (discrBommel et al. [6]), MASON (Luke et al. [41]), MASS package (Tatai et al. [62]), FLAME (Coakley et al. [11], Holcombe et al. [24]), MATSim of EHT Zürich (Bazzan et al. [5]), and SOARS developed in 2010 (Tanuma [60, 61], among others.

ABMs are widely used in modeling of the dynamics of organizations. An example of an agent-oriented model, called the BC model, can be found in the article by Krause [34]. In that model, the agent's attributes include "opinions," and the interactions between agents depend on the distance between their opinions in a non-linear way. These interactions can result in an action being taken by the agent. Other examples of models of social structures based on the concept of opinion interactions can be found in Latane [35] and Galam [21]. A similar approach is presented by Chatterjee and Seneta [7] and Cohen [13]. These works refer to the dynamics of formation of social groups in accordance with the existing agents' attributes (opinions). Some quite interesting results, more closely related to the terrorism problem, are described by Deffuant [18].

Some more general concepts of computational sociology and agent-based modeling can be found in the article of Macy [44]. Other general recommended readings in the field are: Bak et al. [2], Cioffi-Revilla [10], Gotts [23], Axelrod [1], Epstein [19], and Holland [25]. An interesting contribution to a model of the structure of the Osama Bin Laden organization is included in a Vitech Corporation page (link: see References, Long [37]). Other, ABM-oriented approach can be found in Crowder [14] and Hughes [28]. In these publications, we can find discussions about the potential advantages of the ABM approach through a range of examples and through the identification of opportunities in the field of organizational psychology.

Another approach is used by Lustick [43], where the agents interact on a landscape. It is shown that macro-patterns emerge from micro-interactions between agents. An important conclusion is that such effects are more likely when a small number of exclusivist individuals are present in the population. The simulations of other mechanisms of clustering in agent-oriented models are described by Younger [65], who deals with the creation of social structures in the process of food and material storage.

### 7.3.1  People Agents

Human factor always appears in models of social dynamics, economy, interactions of political parties, and groups. Terror organizations [56], self-organizing, self-destruction models [57], and many others. Obviously, constructing models with human actors is a difficult task because of the complexity of the human being and its sometimes irrational behavior.

In the book of Page [47], Chap. 4 (modeling human actor), we can find the term "Rational-Actor Model." In such models, the human behavior is supposed to be governed by individual's preferences when the individual chooses the action that maximizes its income or other benefits. These models work to some extent. Anyway, there is a little choice in defining human behavior. Page illustrates the rational-actor model using the model of person's spending on housing and points out some model imperfections.

Other applications of rational-actor concept appear in models where people maximize their benefit and income. Simulating these mechanisms, we can conclude that the inequality in the society is one of the drivers of the economic growth and welfare. An example of such simulation can be found in [58]. The conclusion from the simulations described in that book is that artificially increasing the equality factor in a society leads to economic disaster and decrease of the overall welfare.

The rational-actor model has limited application, simply because the human behavior, in particular, the macro-behavior of big populations can hardly be considered rational. As stated before, the mass behavior in political activities like mass beliefs and voting activities is, as we can see from the tendencies of the recent tears, in most cases, completely irrational.

Adaptive behavior rules can be used when modeling human factor (Page [47], Chap. 4). The rules change according to the possible payoff an individual can get while taking a specific decision. An example of such behavior is referred to as ait Farol Model in [47].

Other very important mechanism that should be taken into account when simulating human and animal populations is the *herd or Gregarious instinct*. The individuals almost always are looking around and modify their behavior due to the behavior of the others, mainly the near neighbors. The similar model property is called *path dependence* in some sources, like [47]. This mechanism is adopted in the model of self-organization presented in Chap. 8 of the present book.

## 7.4 Discrete Event Specification Formalism (DEVS)

The theoretical base of discrete event simulation is defined in the Discrete Event Specification Formalism (DEVS), see Chow [9].

### 7.4.1 A Remark on Ambiguity

One of the tenets of this book is that the discrete event models and simulation include some elements that may cause some doubts about the validity of the discrete event models. The point is that, if we define the general space of models, the discrete even models may represent a *singularity* in that space. This fact is the cause of some difficulties, mainly while treating possible simultaneous events.

Shortly speaking, the simultaneous events problem can be explained as follows. Recall that in our model and in the simulation program, the simulated *model time* is represented by a model variable that advances while the model events are executed. It is not the *real time* that is the time of our (and of the computer) clocks. Suppose that our computer has one processor that executes the events according to the event messages issued by the event queue mechanism. When two events are scheduled as simultaneous, they must be executed in the same model time instant. However, the computer hardware cannot execute them simultaneously, in the real time. The order of execution, in the real time, must be established by the simulation software or just by the computer hardware. However, the model state, after event execution, may depend on this execution order. So, we get an ambiguity that makes the results software and hardware dependent. Here, we conside one-processor computer.

Consider the following, simple example of the model of a duel (Fig. 7.3).

In this model, we neglect the time interval of several milliseconds, when the projectile travels over the duel area. The events are simply "shoot-and-kill." If both

**Fig. 7.3** A duel

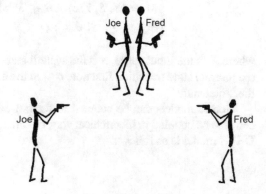

people are perfect shooters, then who shoots first wins, and the other dies. What happens if they shoot simultaneously? A logical conclusion is that the two people die. However, if we simulate this on a computer as simultaneous events, the computer must execute the events in certain order. If the event "Joe shoots" is executed first, then the other event cannot be executed because Fred is already dead. So, the final result will always be "one death" instead of "two deaths." Moreover, the resulting event may also be "Joe dies" so the result is ambigue and hardware dependent. From this example, we can see that the discrete event models may have some inconsistencies, and sometimes may fail. This ambiguity as well as other examples and simultaneous event problems are discussed in Chap. 9.

### 7.4.2  DEVS

There are many real systems, where we can define events that consist in changing the state of the system. For example, the events may describe the start or the end of a service process, a birth or death of a model entity or taking place in a waiting line. In many situations, such events can be considered to be executed in a very small interval of time, compared to the total length of model simulation time. By the *model time* we understand the time variable that is controlled by the simulation program during the simulation run. The *real time* represents the time of our (or computer) physical clock. For example, simulating the movement of a galaxy, we simulate several millions of model time years. On a fast computer, such simulation may take several minutes in the real time. The discrete event simulation means that we suppose that the model events are discrete, i.e., they are accomplished within model time interval of length zero. This model simplification makes the simulations very fast.

The *Discrete Event Specification* (DEVS) formalism is used to describe models in discrete event simulation. In the DEVS formalism, an "atomic" model M is defined as follows (Zeigler [67]):

$$
\begin{cases}
M = \langle X, S, Y, \sigma_{int}, \sigma_{ext}, \lambda, \tau \rangle \\
\sigma_{int} : S \to S, \ \sigma_{ext} : Q \times S \to S, \ \lambda : Q \to Y,
\end{cases}
\tag{7.1}
$$

where $X$ is the input space, $S$ is the system state space, $Y$ is the output space, $\sigma_{int}$ is the internal state transition function, $\sigma_{ext}$ is the external transition function, and $Q$ is the "total state."

Atomic models can be coupled to form a coupled model. The coupled models can also be coupled in hierarchical way, to form more complex models. The coupled DEVS model is as follows:

$$coupled DEVS = \langle X_{self}, Y_{self}, D, \{M\}, \{I\}, \{Z - i, j\}, select \rangle. \qquad (7.2)$$

The sub-index *self* denotes the coupled model itself. $D$ is a set of unique component references. The set of components is

$$\{M_i | i \in D\}. \qquad (7.3)$$

The *select* component defines the order of execution for simultaneous events that may occur in the coupled model. This component must be added to the model to avoid ambiguities in the simulation algorithm and to make the model implementation independent. There is an important research being done on the selected algorithms because the treatment of the simultaneous events is rather difficult task.

The use of the DEVS formalism is relevant in big models, where the time of execution, hierarchical model building, and portability are important factors. To treat complex models with variable structure, the *Dynamic Structure Discrete Event System Specification* (DSDEVS) is used. We will not discuss the DSDEVS formalism here.

The *Time and Event Management* (TEM) includes the clock and event queue management inside the "simulation engine," and the basic queuing model operations provided by the simulation package. The *Object Behavior Modeling* (OBM) is a set of additional items like user-defined distributions and logical functions, non-typical operations, object attributes, and the general object behavior.

## 7.5 Petri Nets

*Petri Nets* (PNs) is a graphical modeling tool for discrete event simulation. A good review of the method can be found in David and Alla [17]. PNs were originally developed by Carl Petri in 1962 to model and analyze communication systems. In PNs there are four elements: *places* (represented by circles), *transitions* (represented by bars), directed arcs, and *tokens* (represented by dots). In simulation terminology, places are conditions, transitions are events, and directed arcs connect places and transitions. A token in a place means that a particular condition holds. The transition fires only when all input conditions are met. If so, it removes a token from each input and deposits a token on each of its output places.

Petri nets are not commonly used, compared to such easy-to-use and quite complete tools like Arena, ProModel, or Simul8. This is an example of an excellent simulation system that remains lost in the huge amount of modeling and simulation paradigms recently available.

Figure 7.4 shows a queuing network and the corresponding Petri net. This is a simple model of a server with its input queue and a network that generates requests

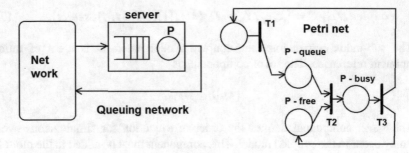

**Fig. 7.4** Input/output validation

to be processed by the server. After processing a request, the server sends a response to the network.

## 7.6 Distributed Simulation Models

The most important way to accelerate the event simulation is to move from sequential to parallel execution. This means that the simulation task is distributed between multiple processors at the same mainframe or in separate computers running in a network.

In distributed simulation, each processor runs with its own model time clock.

Figure 7.5 illustrates problems with time and messages management in distributed simulation. This is a simple model of naval battlefield. The ship and the helicopter

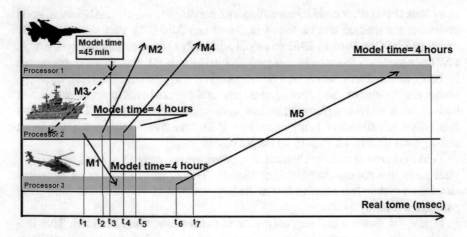

**Fig. 7.5** A naval battlefield

are friendly, and the airplane is the enemy. The simulation tasks of these three components are assigned to three processors, running concurrently. Each of the model components runs with its own clock that shows the model time (in hours). The horizontal axis represents the real computing time, in milliseconds. The local model time clocks are not synchronized. This occurs because the CPU time consuming of the simulation of each component may be different. The most real time consuming is the simulation of the airplane dynamics. The distributed simulation permits to free the processors that have terminated their tasks and dedicate them to other activities. So, the simulation of the movement of the ship terminates more than three times faster than the simulation of the airplane dynamics (for the same final model tine of 4 h).

Model components issue messages to other components or to the environment. Each message has the corresponding *time stamp* that stores the value of the model time, and the action (an order, a piece of information, etc.). The messages are issued in certain instants of the real CPU time $(t_1, t_2, ...., t_7)$.

The message specifications are shown in Table 7.1.

Suppose that the message M3 has not been issued. If so, we can see that the ship issues a message M1 "Eliminate the aircraft" to the helicopter. The helicopter executes the order after some model time, hitting and destroying the airplane, with message M5: "I hit and kill you." The ship sends messages M2 and M4 to the environment (command center).

Now, suppose that the message M3 has been sent at minute 45 of the model time clock of the airplane. If so, the ship must disappear from the battlefield. The ship receives this message in her past. The messages M2, M4, and M5 have already been sent, but they cannot exist because the ship has been eliminated at $t = 45$, and the airplane is not eliminated at 3 h 50 min of model time. This means that the model time clock must be reversed to 45 min, and the simulation must be repeated from this model time instant (the *rollback* in the simulation process). In other words, the clock of the model time of the ship has advanced to more than 3 h, when the plain sent message M3. This means that the message cannot be processed unless the ship goes back with its clock to 45 min of his model time and repeats her simulation from this model time instant. In model with multiple components, such situations result in a chain of rollbacks, and slow down the simulation.

Several distributed simulation techniques have been developed, like Chandy-Misra algorithm Chandy and Holmes [8], and the Time Warp algorithm (Jefferson

**Table 7.1** Message specifications

| Message id | Real time | Model time | Destination | Contents |
|---|---|---|---|---|
| M1 | t1 | 2.5 h | to Helicopter | Eliminate the airplane |
| M2 | t2 | 3.2 h | to External component | My report |
| M3 | t3 | 40 min | to Battleship | I eliminated you |
| M4 | t4 | 3 h 40 min | to External component | Report 2 |
| M5 | t4 | 3 h 50 min | to Battleship | I hit and kill you |

and Sowizral [29]. The former is pessimistic or conservative, advancing the processor simulation clocks only when conditions permit it. In contrast, Time Warp assumes the simulation clocks can be advanced until conflicting information appears; the clocks are then rolled back to a consistent state.

Distributed simulation has no application on single-processor machines and PCs. It is being implemented rather on supercomputers and computer networks. Applications belong mostly to military simulation, communication networks, defense strategy, VLSI chips, and similar great-scale models. Therefore, the fundamental issue is the extent to which simulations should be synchronized in a conservative manner (i.e., without rollback) as opposed to an optimistic manner (i.e., with rollback).

## 7.7　Conclusion

The common, general classification of simulation models refers to *continuous* and *discrete event* simulation. The last kind of models is used in modeling of mass-service systems, manufacturing, and others. This can be done always when we can define the model events as the model state changes that occur in discrete moments of the model time.

The most important theoretical base of discrete event simulation is given in the DEVS (discrete event specification formalism) methodology. In this chapter, we discuss the conventional tools, where the simulation is object- or agent-oriented. Also, some problems related to distributed simulation, like the model time management, are mentioned. The general conclusion from this short overview is that we have a huge number of modeling and software tools, to treat this kind of models. This facilitates the simulation and offers an easy-to-use tool with advanced GUI (graphical user interface). However, while creating the model, one must have certain knowledge about the modeling system. For the models of mass-service systems, it is required that the simulationist is aware of the probabilistic issues of the modeled system.

## 7.8　Questions and Answers

**Question 7.1**　What are the *real time* and *model time*?

**Question 7.2**　What is a *discrete event model*?

**Question 7.3**　What are the *transactions* and *resources* in the GPSS simulation package?

**Question 7.4**　What is the *event message*?

**Question 7.5**　The *event queue* in a discrete event simulation package is defined as follows (select):

1. A queue, created in the computer memory, where the event messages are placed.
2. Waiting line of objects that we want to simulate.
3. Waiting line of the simulation tasks to be executed.
4. Simulated queue of objects, like vehicles on the road or people in a supermarket.

**Question 7.6** Which are the basic strategies in discrete event simulation? Explain.

**Question 7.7** What is the difference between *object-oriented* and *agent-oriented* model?

**Question 7.8** What is DEVS?

**Question 7.9** Why it must be added an additional element *select* to the coupled model?

**Question 7.10** What are *Petri nets*?

**Question 7.11** When the *distributed simulation* can be applied and what are the problems that arise in such kind of simulation?

**Question 7.12** Suppose that entities arrive due to the Poisson input process, to be attended by a server. A waiting line may be formed at the server. What is wrong in the following parameter specification?
   Inter-arrival time interval: Poisson distribution with expected value 5 (minutes).
   Waiting line: FIFO (First in, first out, unlimited).
   Service time (of one entity): Exponential distribution with expected value 6.

**Answers**

**Answer 7.1** The *real time* is the time displayed by our (or computer) clock. The *model time* is one of the model variables. It represents the time advance in the model. For example, we can simulate the evolution of a galaxy over millions of model time years, while the corresponding computer simulation may take several minutes of the real time.

**Answer 7.2** In a *discrete event model*, we assume that the changes of the model state occur in discrete time instants. These changes, called *discrete events*, may take some computing time (the real time), but their duration in the model time is equal to zero.

**Answer 7.3** The GPSS *transactions* are dynamic moving objects like the clients in the bank division, or cars moving over the streets. Transactions appear, move over the model resources, and disappear. The *resources* are fixed model elements.

**Answer 7.4** In a discrete event simulation model, the event message is a data set that must contain the event identifier and the time instant when the event will be executed.

**Answer 7.5**  1. A queue created in the computer memory, where the event messages are placed.

**Answer 7.6**  1. *Activity Scanning* (AS), developed in the 1950s. One of the first implementations was the language CSL. According to this strategy, the model time is set equal to the time instant of the nearest (*time*) event. Then, all model activities (events) are scanned and the time advances to the next event.

2. *Event Scheduling* (ES). In the computer memory, the *event queue* is created. Every component (*event message*) of this queue stores the time when the event will be executed, and the event identifier.

3. *Process Interaction* (PI). The model is defined in terms of processes that can run concurrently. The rules of interaction between processes are defined, while the specification of a process includes the necessary event scheduling. PI can be implemented in any object-oriented programming language.

**Answer 7.7**  In the discrete, object-based modeling, we create *objects* that behave according to the user-defined rules and execute their events in discrete moments of the model time. The agent-based models manage objects called *agents*, which are equipped with certain "intelligence." They can take decisions, optimize their behavior, and obey some more complicated behavior rules.

**Answer 7.8**  DEVS (Discrete Event Specification Formalism) is a theoretical base for discrete event modeling and simulation. In DEVS, we can define an elemental, "atomic" models, and then use them to create *coupled models*. These models can be coupled again, in a hierarchical way, to create big, complex models. DEVS models support portability and reuse of sub-models. A considerable speed-up of the simulations is achieved.

**Answer 7.9**  The *select* element is added to handle simultaneous events. Such events may provoke an ambiguity in the simulations. *Select* element resolves potential conflicts and ambiguity if simultaneous events occur.

**Answer 7.10**  *Petri nets* (PNs) is a modeling tool for discrete event simulation. PN model includes four elements: *places* (represented by circles), *transitions* (represented by bars), directed arcs, and *tokens* (represented by dots). In simulation terminology, places are conditions, transitions are events, and directed arcs connect places and transitions. A *token* in a place means that the necessary conditions hold. The transition fires (the event is executed) only when all input conditions are met. If so, it removes a token from each input and deposits a token on each of its output places.

**Answer 7.11**  *Distributed simulation* can be run on computers with multiprocessing (parallel event execution on multiple processors).

To accelerate the simulation, the tasks that run concurrently have their own model time clocks. In general, the concurrent tasks are not synchronized because some of them may be very fast, and others need more (real) computer time to be completed. When the concurrent tasks send messages and interact with each other, this lack of synchronization may provoke problems in the management of the local model time clocks.

**Answer 7.12** The inter-arrival interval for the Poisson arrival process cannot have the Poisson distribution. This is perhaps the most common and "fatal" error in discrete event simulation. The inter-arrival interval for this process has the *exponential* distribution.

# References

1. Axelrod R (1997) The dissemination of culture - a model with local convergence and global polarization. J Conflict Resolut 41(2):203–226
2. Bak P, Paczuski M, Shubik M (1997) Price variations in a stock market with many agents. Physica A: Stat Mech Appl 246(3):430–453
3. Bandini S, Manzoni S, Vizzari G (2009) Agent based modeling and simulation: an informatics perspective. J Artif Soc Soc Simul 12(4)
4. Baron RA (1977) Aggression: definitions and perspectives. In: Human aggression. Perspectives in social psychology (A series of texts and monographs). Springer, Boston, MA. https://doi.org/10.1007/978-1-4615-7195-7_1
5. Bazzan AIC (2009) Auton Agents Multi-Agent Syst 18(3):342–375. https://doi.org/10.1007/s10458-008-9062-9
6. Bommel P, Becu N, Le Page C, Bousquet F (2015) Cormas, an agent-based simulation platform for coupling human decisions with computerized dynamics. In: Hybrid simulation and gaming in the network society series. Translational Systems Sciences. Springer, Singapore. https://doi.org/10.1007/978-981-10-0575-6_27
7. Chatterjee S, Seneta E (1977) Towards consensus: some convergence theorems on repeated averaging. J Appl Probab 14(1):89–97
8. Chandy KM, Holmes V, Misra J (1979) Distributed simulation of networks. Comput Netw 3(1):105–113
9. Chow AC, Zeigler BP (1994) The simulators of the Parallel DEVS formalism. In: Proceedings of the fifth annual conference on AI, simulation and planning in high autonomy systems
10. Cioffi-Revilla C (1998) Politics and uncertainty: theory, models and applications. Cambridge University Press, Cambridge
11. Coakley S, Smallwood R, Holcombe M (2006) From molecules to insect communities – how formal agent based computational modeling is undercovering new biological facts. http://www.jams.or.jp/scm/contents/e-2006-7/2006-69.pdf. Scientiae Mathematicae Japonicae Online, e-2006, 765–778
12. Coe RM (1964) Conflict, interference, and aggression. Behav Sci 186–197
13. Cohen JE, Hajnal J, Newman CM (1986) Approaching consensus can be delicate when positions harden. Stoch Process Appl 22(2):315–322
14. Crowder RM, Robinson MA, Hughes HPN, Sim YW (2012) The development of an agent-based modeling framework for simulating engineering team work. IEEE Trans Syst Man Cybern Part A Syst 42(6):1426–1439
15. Dahl O, Nygaard B (1967) Simula - an Algol-based simulation language. Commun ACM 9:671–678
16. Danaf M, Abou-Zeid M, Kaysi I (2015) Modeling anger and aggressive driving behavior in a dynamic choice–latent variable model. Accid Anal Prev 75:105–118. ISSN 0001-4575, https://doi.org/10.1016/j.aap.2014.11.012
17. David R, Alla H. Petri nets for modeling of dynamic systems: a survey. Automatica 30(2), Elsevier. https://doi.org/10.1016/0005-1098(94)90024-8
18. Deffuant G, Amblard F, Weisbuch G, Faure T (2002) How can extremism prevail? A study based on the relative agreement interaction model. J Artif Soc Soc Simul 5(4)
19. Epstein JM, Axtell R (1996) Growing artificial societies: social science from the bottom up Brookings Institution Press, Washington, DC

20. Forrester JW (1958) Industrial dynamics-a major breakthrough for decision makers. Harvard Bus Rev 36(4):37–66
21. Galam S, Wonczak S (2000) Dictatorship from majority rule voting. Euro Phys J B 18(1):183–186
22. Gordon G (1975) The application of GPSS to discrete system simulation. Prentice-Hall, Englewood Cliffs
23. Gotts NM, Polhill JG, Law ANR (2003) Agent-based simulation in the study of social dilemmas. Artif Intell Rev 9(1):3–92
24. Holcombe M, Coakley S, Kiran M (2013) Large-scale modelling of economic systems. Compl Syst 22(2):175–191. http://www.complex-systems.com/pdf/22-2-3.pdf
25. Holland JH (1998) Emergence: from chaos to order. Addison-Wesley Publishing Company, Helix Books
26. Hauge J, Paige K (2001) Learning SIMUL8: the complete guide (and SIMUL8 Version 6). PlainVu Publishers (and SIMUL8 Corporation), Bellingham, WA
27. Hollocks B (2006) Forty years of discrete-event simulation—a personal reflection. J Oper Res Soc 57:1383–1399. https://doi.org/10.1057/palgrave.jors.2602128
28. Hughes HPN, Clegg CW, Robinson MA, Crowder RM (2012) Agent-based modelling and simulation: the potential contribution to organizational psychology. J Occup Organ Psychol 85:487–502
29. Jefferson D, Sowizral H (1985) Fast concurrent simulation using the Time Warp mechanism, Distributed Simulation 1985, The 1985 Society of Computer Simulation Multiconference. San Diego, California
30. Kadar DZ, Parvaresh V, Ning P (2019) Mortality, morel order, and language conflict aggression. J Lang Aggress Confl 7(1):6–31. https://doi.org/10.1075/jlac.00017.kad
31. Kelton D, Sadowski R, Sadowski D (2004) Simulation with ARENA. McGraw-Hill, New York
32. Kim SS (1976) The Lorenzian theory of aggression and peace research: a critique. J Peace Res 13(4):253–276. https://doi.org/10.1177/002234337601300401
33. Klein J (2002) Breve: a 3D environment for the simulation of decentralized systems and artificial life. In: Conference paper: ICAL 2003 proceedings of the eighth international conference on Artificial life, MIT Press, Cambridge, MA. ISBN/ISSN 0-262-69281-3
34. Krause U (2000) A discrete nonlinear and non-autonomous model of consensus formation. In: Elaydi S, Ladas G, Popenda J, Rakowski (eds) Communications in difference equations. Gordon and Breach, Amsterdam
35. Latane B, Nowak A (1997) Self-organizing social systems: necessary and sufficient conditions for the emergence of clustering, consolidation and continuing diversity. In: Barnett FJ, Boster FJ (eds) Progress in communication sciences v.13. Ablex Publishing Corporation. ISBN-13: 978-1567502770
36. Lin JT, Lee CC (1993) A three-phase discrete event simulation with EPNSim graphs. SIMULATION. 60(6):382–392. https://doi.org/10.1177/003754979306000603
37. Long JE (2002) Systems analysis: a tool to understand and predict terrorist activities. Internet communication Vitech Corporation. http://www.umsl.edu/~sauterv/analysis/62S-Long-INTEL.pdf
38. Lorenz KZ (1964) Ritualized fighting. In: Carthy JD, Ebling FK Jr (eds) The outward handling of aggression. Academic Press, N.Y.
39. Lorenz KZ (2002) On aggression. Psychology Press
40. Lorenz KZ (1981) The centrally coordinated movement or fixed motor pattern. In: The foundations of ethology. Springer, Vienna. https://doi.org/10.1007/978-3-7091-3671-3-6
41. Luke S, Cioffi-Revilla C, Panait L, Sullivan K (2005) MASON: a multiagent simulation environment. Simulation 81(7):517–527
42. Luo D (1997) Bifurcation theory and methods of dynamical systems, p 26. World Scientific. ISBN 981-02-2094-4
43. Lustick S (2000) Agent-based modeling of collective identity. J Artif Soc Soc Simul 3(1). http://jasss.soc.surrey.ac.uk/3/1/1.html

44. Macy MW, Willer R (2002) From factors to actors: computational sociology and agent-based modeling. Annu Rev Sociol 28(1):143–166

45. Michael JN, Nicholson T, Collier JR, Vos JR (2006) Experiences creating three implementations of the repast agent modeling toolkit. ACM Trans Model Comput Simul 16(1):1–25. https://doi.org/10.1145/1122012.1122013

46. O'Keefe RM (1986) The three-phase approach: a comment on strategy-related characteristics of discrete event languages and models. Simulation 47(5):208–210

47. Page SE (2018) The model thinker. Hachette Book Group, New York. ISBN 978-0-465-009462-2

48. Parker MT (2001) What is ascape and why should you care? J Artif Soc Soc Simul. http://jasss.soc.surrey.ac.uk/4/1/5.html

49. Pegden CD, Sturrok DT (2010) Introduction to Simio. In: Conference paper: proceedings of the 2010 Winter, PA, USA

50. Pegden D, Ham I (1982) Simulation of manufacturing systems using SIMAN. CIRP Ann 31(1):365–369. ISSN 0007-8506, https://doi.org/10.1016/S0007-8506(07)63329-0

51. Pritsker A (1986) Introduction to simulation and SLAM II. Wiley & Sons, Inc. ISBN: 978-0-470-20087-2

52. Railsback SF, Lytinen SL, Jackson SK (2006) Agent-based simulation platforms: review. Simulation 82(9):609–623. https://doi.org/10.1177/0037549706073695

53. Raczynski S (2003) Continuous simulation. In: Encyclopedia of information systems. Academic Press, Elsevier Publ, New York, NY

54. Raczynski S (2009) Discrete event approach to the classical system dynamics. In: Conference paper: Huntsville simulation conference, SCS, Huntsville AL

55. Raczynski S (2019) Interacting complexities of herds and social organizations: agent based modeling. Springer

56. Raczynski S (2004) Simulation of the dynamic interactions between terror and anti-terror organizational structures. J Artif Soc Soc Simul 7(2). England. http://jasss.soc.surrey.ac.uk/7/2/8.html

57. Raczynski S (2006) A self-destruction game. J Nonlinear Dyn Psychol Life Sci 471–483

58. Raczynski S (2021) Catastrophes and unexpected behavior patterns in complex artificial populations. Springer. ISSN 2198-2404

59. SWARM Development Group (2001) Swarm simulation system. Electronic citation. Electron 8(1–10). http://digitalcommons.usu.edu/nrei/vol8/iss1/2

60. Tanuma H, Deguchi H, Shimizu T (2005) Agent-based simulation: from modeling methodologies to real-world applications, vol 1. Springer, Tokyo

61. Tanuma H, Deguchi H, Shimizu T (2006) SOARS: spot oriented agent role simulator – design and implementation. In: Agent-based simulation: from modeling methodologies to real-world applications. Springer, Tokyo. ISBN 9784431269250

62. Tatai G, Gulyas L, Laufer L, Ivanyi M (2005) Artificial agents helping to stock up on knowledge. In: Conference paper: 4th international central and eastern european conference on multi-agent system, Budapest, Hungary, ISBN: 3-540-29046-X 978-3-540-29046-9. https://doi.org/10.1007/11559221_3

63. Thom R (1975) Structural stability and morphogenesis (D.H. Fowler, Trans.). Benjamin-Addison Wesley, New York

64. Weaver DS (1980) Catastrophe theory and human evolution. J Anthropol Res 36(4):403-10. Accessed Jan 8, 2021. http://www.jstor.org/stable/3629609

65. Younger SM (2003) Discrete agent simulations of the effect of simple social structures on the benefits of resource. J Artif Soc Soc Simul 6(3)

66. Zeeman EC (1976) Catastrophe theory. Sci Am 234(4):65–83. https://doi.org/10.1038/scientificamerican0476-65

67. Zeigler B (1987) Hierarchical, modular discrete-event modelling in an object-oriented environment. Simulation 49(5):219–230. https://doi.org/10.1177/003754978704900506

# Chapter 8
# Self-Organization, Organization Dynamics, and Agent-Based Model

## 8.1 Introduction

A model of the dynamics and interactions between organizations with self-organizing hierarchical structures is presented. The active objects of the model are individuals (people, organization members). The parameters of an individual are ability, corruption level, resources, and lust for power, among others. Three organizations are generated and interact with each other, attempting to gain more members and power. The individuals appear and disappear, due to a simple "birth-and-death" process. If an individual disappears from the model, then the corresponding reconfiguration in the hierarchical structure takes place. The growth of organizations and macro-patterns is the result of the activities of the individuals. The aim of the simulation is to visualize the evolution of the organizations and the stability of the whole system. A "steady state" for the model is hardly achieved. Instead, in most parameter configurations, the model enters into oscillations.

The model presented here is an abstract one, not related to any real social or political system. So, the results should be only treated as qualitative. However, these qualitative results may provide hits and a better understanding of real organizations. The individuals should be understood in a more general sense. These may be people, but also groups of people, as well as sub-organizations. So, we will use a more generic term entity, used in the literature on the discrete event models (commonly referred to as DEV or DEVS in the literature). The model entities, as well as the simulated organizations, are charged with a certain corruption level, as explained forthwith. Thus, this kind of model depicts political, business organizations, and trade unions, rather than welfare or benevolent institutions.

It is known that the main goal of any political party is to obtain power and nothing more. Many trade union organizations have lost sight of their original goal (defending the interests of workers) and have also become power-seeking structures. A social structure acts as a new agent, using its members as nothing more than a medium to achieve its goal. However, in this model, an organization itself is not an active process. The organization macro-patterns are the results of the activities of its members.

© The Author(s), under exclusive license to Springer Nature Switzerland AG 2022     189
S. Raczynski, *Models for Research and Understanding*, Simulation Foundations, Methods and Applications, https://doi.org/10.1007/978-3-031-11926-2_8

The interaction between different social structures can be simulated, to some extent of course, as described in Raczynski [31] (the simulation of interactions between terrorist and anti-terrorist structures.) Here, a similar approach and tools are used.

Many existing models of the dynamics of social organizations use the *agent-based modeling* (ABM). An interesting agent-oriented model, called the BC model, can be found in the article by Krause [23]. In that model, the agent's attributes include "opinions," and the interaction between agents depends on the distance between their opinions in a non-linear way. These interactions can result in an action being taken by the agent. Other examples of models of social structures based on the concept of opinion interactions can be found in Latane [25], and Galam [19]. A similar approach is presented by Chatterjee [7] and Cohen [11]. The BC model and the above works refer to the dynamics of forming of social groups in accordance with the existing agents' attributes (opinions). Some quite interesting results, more closely related to the terrorism problem, are described by Deffuant [14].

Another agent-oriented approach is used by Lustick [27], where the agents interact on a landscape. It is shown that macro-patterns emerge from micro-interactions between agents. An interesting conclusion is that such effects are more likely when a small number of exclusivist identities are present in the population. The simulations of other mechanisms of clustering in agent-oriented models are described by Younger [41], who deals with the creation of social structures in the process of food and material storage.

Some more general concepts of "computational sociology" and ABM modeling can be found in the article of Macy [28]. Other general recommended readings in the field are: Bak [4], Cioffi-Revilla [10], Gotts [20], Axelrod [3], Epstein [18], and Holland [21]. A model of the structure of the Osama bin Laden organization, included in a Vitech Corporation page (link: see References, Long, [26]) is also an interesting contribution. Other, (ABM)-oriented approach can be found in Crowder [13] and Hughes [22]. In these publications, we can find notes on the potential advantages of the ABM approach through a range of examples and through the identification of opportunities in the field of organizational psychology.

A very basic and comprehensive text on the organization theory and dynamics can be found in Daft [12]. The book contains classic ideas and theories, and real world practice. The problems and questions addressed are: "How organizations adapt to, or control competitors, customers, government, and the environment? How to avoid management ethical lapses? How to cope growing bureaucracy? How to manage the use of power and politics among managers? What structural changes are needed?" among others. Throughout the text, detailed examples illustrate how companies behave in the rapidly changing, highly competitive, international environment. However, these topics are quite different from what we consider in this chapter. Daft does not consider modeling and simulation as an important tool in organizational design. The works like that address the organizational theory just from other, perhaps more practical, perspective. Organization theory and design are treated in a huge number of publications, sometimes from a completely different point of view.

It should be noted that looking for a model that simulates real human behavior is Utopian. Nobody had ever simulated a human in its complete (mental, emotional, physical, etc.) behavior. All that can be done is to choose some little part of this complex system in order to simulate its possible actions. In any case, in soft system simulation and social simulation, one can hardly ever (or never) find any proof that the model is valid.

Interesting models and simulation experiments on the survival of societies can be found in the literature. Cecconi [9] simulates a survival problem in terms of individual or social resources storage strategies. Saam [34] simulates the problems of social norms, social behavior, and aggression in relation to social inequality. Staller et el. [39] discuss the emotional factor in social modeling. They introduce the emotions as an essential element of models that simulate social behaviors. Stocker [40] examines the stability of random social network structures in which the opinions of individuals change. It is pointed out that hierarchies with few layers are more likely to be more unstable than deeper hierarchies. The study is related to political, organizational, social and educational contexts rather than to the destruction problem itself, but it is clear that an unstable social structure may be much more vulnerable to attack. There are many approaches and aspects of ecological and social models, providing certain reproduction/death formulas. See, for example, Moss de Oliveira [29], for a model of aging and reproduction. The problem of survival and self-destruction treated from the ABS framework can also be found in Raczynski [32].

Adamic et al. [1] address the question of how participants in a small world experiment are able to find short paths in a social network using only local information about their immediate contacts. A contact's position in physical space or in an organizational hierarchy relative to the target is considered. The authors discuss the implications of their research in social software design.

From a newer publications, we should mention the book of Edmonds et al. [15] that is a collection of interesting papers. The editors aimed to present a flyover of the current state of the art. They divide the 24 papers into three parts: model oriented, empirically oriented, and experimentally oriented. In the other publication of Edmonds [16] we can find an analysis of the role and effects of context on social simulation.

Silverman [38], presents an agent-based model of a human population. The model illustrates the potential synergies between demography and agent-based social simulation. Elsenbroich [17] asks what kind of knowledge can we obtain from agent-based models. The author defends agent-based modeling against a recent criticism. Sibertin-Blanc [37] presents a framework for the modeling, the simulation, and the analysis of power relationships in social organizations and more generally, in systems of organized action. In that article, we can find a discussion of a model of bounded-rational social actors and analytical tools for the study of the internal properties of organizations. The model pretends to explain why, in an organizational context, people behave as they do.

The agent-based modeling is a powerful tool, very different from other modeling paradigms, mainly Systems Dynamics (SD). In SD we start from the interaction rules for the model global variables and from the structure of the real system to generate the system trajectories. The ABM approach is quite opposite: the interactions between the variables are unknown, and the model is constructed by defining the events that may occur in the "life" of model components, named agents. Some artificial intelligence, like the ability to take decisions and to interact with other agents can be added to the agent specification. However, no interactions between global variables, like the size of the organizations, are known. The global behavior of the model, the trajectories of the model variables, and their eventual relations are the results of the simulation. In other words, the agents are running their events, which results in the model macro behavior. They form a system, in which behavior is not just a sum of the actions of individual components. This is a known property of complex systems, related also to the property of non-linearity (see Schachter [35]). No differential equations are defined or used, like in the SD. This is the great advantage of ABM simulation, because not all that occurs in the real system is governed by the differential equations (something difficult to understand by electrical engineers).

An exhaustive comparison between SD and ABM has been done by Borshchev and Filippov [5]. An interesting suggestion included in their paper is that the ABM can be used as an add-on, which can be efficiently combined with System Dynamics and Discrete Event modeling, resulting in multi-paradigm model architectures. An attempt to create an SD-like simulation tool based on ABM and not on the differential equations is presented in Raczynski [33].

## 8.2   The Model

Our model is rather abstract and can hardly be validated for real organization in a quantitative sense. However, a qualitative comparison with real organization dynamics may be done. For example, the oscillatory pattern of the size of real competing political parties coincides with the results of our model. The qualitative results provided by the model can be used as hints for the predictions about the real system behavior. Note that the members of the model organizations move over a political map PM we introduce here. This map is a multi-dimensional "space of ideas," which coordinates may represent, for example, the level of "democratic orientation," "totalitarianism" or "religious orthodoxy" of the moving entities. For a simpler models this PM can be interpreted as a geographical landscape where the agents move.

The concept of *corruption* in this chapter should be interpreted in the very general terms. It may be unethical/illegal behavior, or just a deterioration of certain ideological patterns or opinions. So, the corruption level can be also associated with a spot in the political map. The main assumption is that corrupted spots on the political map provide little benefit to the model entities. Thus, new entities that appear on the map tend to avoid these spots. The corruption is also a topic of some interesting publications. Anand and Blake [2] consider rationalizations, which are mental strategies that allow employees (and others around them) to view their corrupt acts as justified. Another approach to corruption in organizations can be found in Pinto [30] or Lambsdorff [24]. However, most of the academic papers on this subject are based on historic data analysis or psychological and social issues, rather than modeling and computer simulation.

An interesting, quantitative approach to the concept of corruption can be found in Caulkins et al. [6], related to the earlier work of Shelling [36]. The authors are looking for a "stable equilibrium levels of corruption" in their model. The point of equilibrium is found as a solution to an optimization problem. The decision-makers or leaders are supposed to follow the solution to an infinite-time non-linear optimal control problem. The model is continuous, and its dynamics is described by ordinary differential equations. However, as stated before, in the real world, and in particular in the dynamics of organizations with human factor, the variables hardly obey differential equations, and sometimes even a simple logic. So, the ABM models, where the only thing we define are possible events in the most elemental model components (entities, members of the organization), seem to be more realistic. As for a possible point of equilibrium, it is rather questionable if such point can be reached. The real organizations are in constant movement and hardly can rest in a theoretical "equilibrium point."

As the model is not related to any real organizations and provides qualitative results only, it cannot be strictly validated, for example through the input-output validation. However, taking into account the qualitative model behavior, and looking, for example, at the size of the Republican and Democratic parties in the US, we can see that the main model property (oscillatory behavior) is very similar to the reality.

The main point of this chapter is that the ABM modeling can provide interesting hints on organizational dynamics. The model trajectories show that no static equilibrium of the model is reached. The resulting model movement can be interpreted as the orbital stability known from the control theory, see Weinstein [42], Chen [8]. However, remember that no differential equations are used to describe the dynamics. So, the concepts of control theory, like stability and optimality cannot be used here directly as done by Caulkins [6].

Our model consists of three hierarchical structures interacting with each other over a common (abstract) region. Let us comment on some terms used here.

**Entity**. An individual that can be a member of a hierarchical structure.

**Organization**. A collection of entities, with a hierarchical structure. In this simulation, no initial structure is imposed on the organizations. They are self-organizing, starting from the "chaos" (chaotic set of entities). Each organization has a corruption parameter, telling how corrupt or "spoiled" the organization is. The corruption level is calculated as the weighted average of the corruption parameters of all its members. The weight is equal to the reciprocal of the entity level in the organization. The head of the organization has level 1, its subordinates have level 2, 3... etc.

**Political Map (PM)**. This is one- or multi-dimensional region, where the entities are placed. The PM should be understood in very general terms. It can be just a geographical region, or a generalized space of ideas or political orientation. For example, in a two-dimensional case, one axis may be a religious orientation (from *atheism* to *religious extremism*), and the other may be the ideology (from *democracy* to *totalitarianism or communism*).

**PM Corruption Field (CF)**. One of the concepts implemented here, related to the PM is the assumption that the political and social ideas are subject to wear. What was supposed to be a good idea a hundred years ago, is hardly considered good now, due to the corrupted organizations that resulted from its implementation. The CF is a function of the spatial variable (entity's position on the PM), that tells how "good" the spot is. It returns zero if the spot is completely spoiled and one if it is a good spot. The value of CF is used by the entities that appear (are born, created) on the PM. The higher the CF is, the higher is the probability that the new entity occupies the place. In other words, the CF defines the probability distribution for the coordinates of new entities. The value of CF on the spot on the PM may be interpreted in many ways. It may be the ideological deterioration (obsolete and erroneous trends and beliefs) or just a position that, after some time, no longer provides incentives and benefits to the entity.

**Time**. The model time is measured in abstract time units (TU). The simulations are run with a final simulation time equal to 2000 or 5000 TU.

**Entity personal data**. It is the collection of the following parameters:

*Ability*. This is just the ability to climb in the hierarchy of the organization. Note that such concepts as *intelligence or education* do not exist in this model, being irrelevant in politics.

*Lust for power*. This is the most important entity parameter. In other words, the entity may become a leader if it really wants to, which occurs in the real political life.

*Resources*. The financial or other resources that help the entity to climb in the hierarchy.

*Corruption level*. This parameter takes values from honest to totally corrupt. As stated before, the corruption level can be the result of the unethical/illegal behavior or other causes, like the rationalization tactics used by individuals committing unethical or fraudulent acts. The corruption level may also be interpreted in a more general sense. For example, it may be degenerative changes in the mentality of individuals that occupy, for a long time, a high position in the organization (mainly political).

*PM coordinates*. The place the entity takes on the PM. In general, it is the entity political orientation. As stated before, this may be a scalar or a set of coordinates on multi-dimensional PM. In this simulation experiments, the PM is two-dimensional (mostly for the sake of image clarity) and its image on the screen is a square.

*Life time*. The lifetime determines when the entity dies or just disappears from PM (natural death). Lifetime is defined as a random variable with density function $\exp(70.0)$,

*Superior*. The pointer to another entity, the "boss." The entity is one of the subordinates of the boss.

*Subordinates*. Pointers to the subordinates of the entity. Each entity can have any number of subordinates. However, for the sake of clarity in the organization images, it is supposed that the entity should have four subordinates. So, if the number of subordinates is less than 4, the entity attempts to catch more subordinates.

No physical units for the ability, lust of power, resources, and the corruption level are defined. All these parameters are relative, with values in [0,1].

## 8.2.1 Interaction Rules

It should be noted that using ABM the only model specification we need are the individual interaction rules, defined by the events the entities execute during their "life." The model instability (or rather orbital stability) is the result of the individual actions of the entities. There are no global rules: the entities are being launched and what we obtain is the result of their individual actions. The resulting structure of the organizations is the result of self-organizing. An organization is just a data structure and does not take any actions of its own. However, organizations behave as if they had a specific goal: to grow and keep growing.

The simulation program has been coded using the *Bluesss Simulation System*. Recall that the main concepts of Bluesss are processes and events (see Sect. 8.3). A *process* is a template, like a class declaration in object-oriented languages. At the runtime objects (entities) are generated, being instants of the process declaration. Within a process, a several *events* are declared. The event execution is controlled by Bluesss package, according to the clock mechanism and to the internal event queue. For more detail consult http://www.raczynski.com/pn/bluesss.htm .

The model includes two processes: **entity** and **monitor**. Note that the "organization" is not represented by any particular process; it is just a data structure. So, the organization itself has no "awareness" and does not take any actions. On the other hand, for an external observer, organizations behave as systems with their own goals (to grow and to gain power). The evolution of the organization parameters is the result of the simulation. The only rules of interaction are those of the members.

Model entities are created by the *monitor process*. After being created, the entity occupies place on the PM, due to a simple rule: the higher is the corruption level on the spot, the lower is the probability the entity will appear there. The monitor also initializes three organizations, marking three (randomly chosen) entities as organization heads. In this chapter, the growing organizations have a simple hierarchical structure; each entity can have several subordinates which, in turn, have their own subordinates and so on.

The interaction rules are defined by the actions taken by the entities, defined by the following events:

**Seek for subordinates**. At the very beginning, only the organization top entities (heads) seek for subordinates. This is done repeatedly, until the entity has gained four subordinates. Each subordinate starts to seek for their subordinates, and so on. Any entity that is the head of the organization or has its superior and less than four subordinates does it.

**Die**. This makes the entity disappear from the PM. The event occurs at the end of the entity life time. If the entity was a member of an organization, then one of its subordinates (say X, if any) takes its place. A subordinate of X takes the place of X and so on, iteratively. Another possibility of the entity disappearance is the result of an action of one of its subordinates (the climb event).

**Climb**. The entity makes disappear his superior and takes its place. A subordinate of the entity takes its place and so on, iteratively. To be able to climb, the sum of the entity lust for power, ability, and resources must be greater than the same sum of its superior. As the entity superior may change, this attempt is repeated every 30 TU, in average (exponential distribution).

**Move**. This is a slow random walk of the entity over the PM. The entity changes randomly its position by a small amount. The event is repeated every TU. There are also two modes of additional movement:

*Attract mode*: the entity is being attracted by the head of the organization it belongs, and by its boss.

*Escape mode*: the entity escapes toward the less spoiled spots.

These two modes are enabled by the logical variables *attract* and *escape*, respectively.

**Propagate**. The head of each organization propagates his own corruption level to all members of the organization. Each entity changes its corruption level as follows:

*entitycorruptionlevel = 0.1\*headcorruptionlevel + 0.9\*entitycorruptionlevel*

This event is repeated each time unit. So, the corruption parameter within the organization becomes more uniform.

**Modify PM**. The entity changes the local value of the corruption field CF. The whole PM region is divided into 900 (30x30) square elements, each of them with its corresponding CF value. In this event, a factor value is calculated using the following formula:

*F = (corruptionlevel/level + orgcorr)\*0.04,*

Where *corruptionlevel* and *level* are parameters of the current entity, and *orgcorr* is the corruption level of the organization it belongs to. So, the entities with lower level value have less influence on the CF value (the head level is equal to 1, its subordinates have one level 2, and so on). The entity repeats this event each 0.5-time units.

In such way, some parts of the PM become corrupted. The value of the CF is truncated to [0,1]. On the other hand, the CF recuperates constantly. The monitor process augments the CF in each spot by 0.015, each time unit. All this makes the CF change constantly, depending on how corrupt is the organization that occupies the spot. Recall that CF, after being normalized, is used as the probability density function for the appearance of new entities. Note that this recuperation process is in some sense the reflection of the lack of historical memory in the real societies. It is well-known fact that the societies learn nothing from their history. Thus, the ideas that failed some time ago, become attractive again after some time.

The agent parameters are not taken from any real organization, and the model is rather abstract. However, observe that the above actions of the agents are similar to the activities of members of many real organizations.

## 8.3   BLUESSS Simulation Package

The Bluesss package (Blues Simulation System) was used to simulate the model. Let us recall the main features of the package. The main features of the package are as follows:

Object-oriented simulation

Continuous/discrete models

Clock mechanism

Inheritance

Easy to use, clear process/event structure

Low cost

BLUESSS runs with the Embarcadero C++Builder. As the Bluesss code is translated to C++, the package is extremely flexible; it can use all the features available from the C++ code.

The Bluesss source code is converted into C++, then compiled and executed. The resulted exe file is a stand-alone Windows application. For more information, consult http://www.raczynski.com/pn/bluesss.htm. The general concept is to declare a series of processes and, inside each process several events that can occur during the "life" of the corresponding object. The objects are the instances of process declarations. In the following, by a *process* we understand an instance (object) created according to the corresponding *process declaration*. For example, if we declare a process "client" as a client at a bank, then we can generate and launch thousands of clients, each of them running its own events, such as "enter the bank," "join a waiting line," and "occupy a teller." Each event has a body that may include a code for any, simple or complicated, algorithm. Therefore, the user creates objects that represent the model agents and equip them with a desired behavior. This makes Bluesss a good tool for agent-based modeling. Inside the process declaration, there are several event definitions. The events can be scheduled for their execution, which is controlled by the Bluesss clock mechanism.

As the simulation process passes through the C++ compiler, we can use any feature that is available from C++Builder. Each object can interact with other objects, change its own attributes or those of other objects, execute a complicated computational procedure, execute an external program, display an image, sing a song, or communicate by the internet. These are features needed in agent-based simulation. Such actions as making decisions about where and when to migrate, or following the crowd (gregarious) instinct can be simulated. To run models like that described in this paper, any of the software tools mentioned in the previous section can be used. However, we used of Bluesss not only because this is a software developed by the author. This tool is perhaps not such user-friendly as other packages because it requires some abilities in C++ programming. On the other hand, it is extremely flexible and permits the use of anything available from C++. This way, we can insert into the events executed by the generated entities a simple, as well as complicated behavior algorithms. There is a similar tool named PSM++, based on the Delphi package. The only difference is that in PSM++ the code of the event body if written in Delphi Pascal instead of C++. Some examples of PSM++ applications in artificial society simulation can be found in [31, 32].

## 8.4   Simulations

At the beginning of the simulation run, one instance of the *monitor* process is created and activated. It creates 1000 entities randomly located over the PM. For each entity, its parameters are being defined and the events seek for subordinates, move, modify PM and climb are invoked. The entity event *die* is scheduled to be executed at the actual model time (when the entity was created) plus the entity lifetime. If the entity has disappeared earlier, this event is ignored. In the monitor process, the necessary events are initialized, such as initiating organizations (mark the heading entities) organization state display, and CF recovery. The monitor process also stores the model state parameters for further analysis and trajectory plotting. Then, all other events are executed automatically. The organizations grow, entities move and execute their own events. Figure 8.1 shows a typical image of the PM at the initial stage (growing organizations).

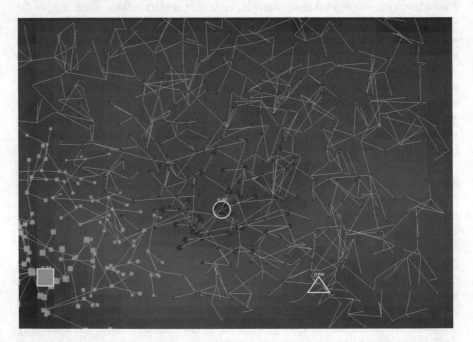

**Fig. 8.1**  Initial simulation stage: The PM with growing organizations

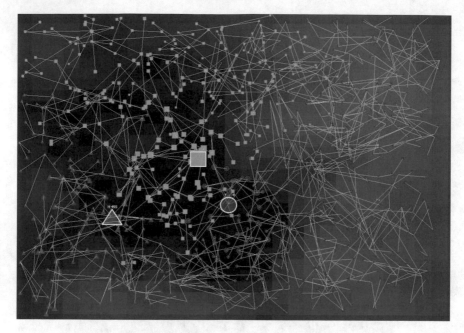

**Fig. 8.2** The PM after 500 model time units

The three organizations are marked with circles, squares and triangles, respectively. Small gray points represent new entities, not affiliated yet. The lines are links between superior and subordinate. The big figure is the organization head, and the size of the figures decreases for entities with descending level inside the organization. The situation after about 500-time units is shown in Fig. 8.2. The "good" spots are shown as light blue, and the spoiled areas are in dark.

The monitor process shows the situation on the PM with small time steps, providing an animated image. It is a nice program feature, where the entities move over the area and the "spoiled" and "good" PM regions change color.

As stated before, the experiments provide only a qualitative information. The model behavior is not easy to predict from the specifications of the model components and interaction rules. There are some possible scenarios. One would expect that the size of the organizations as well as the other variables will change chaotically. Another possibility is that one or two organizations will collapse and, after a long simulation time, only one, the strongest "winning" organization will remain. we also may suppose that the model reaches some kind of equilibrium. The experiments show that none of the above occurs. After a short initial transitory period, the model enters in quite regular oscillations. No "steady state" is reached. Figure 8.3 shows the size of the three organizations, in relation to the size of the whole population. The shape of the curves resembles interference between signals with slightly different

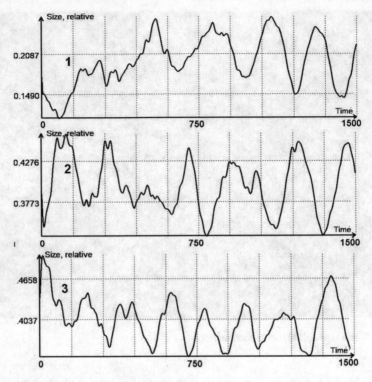

**Fig. 8.3** Relative size of organization 1, 2, and 3

frequencies. In our model everything is stochastic, so each simulation is different. However, this oscillatory nature of the model can always be observed. Recalling concepts of stability of the control theory, the model seems to be orbitally stable, see Chen [8], Weinstein [42].

The attraction mode seems to stabilize the organizations. Figure 8.4 shows the image of the organizations after 500 time-units. In Fig. 8.5 we can see the changes in the relative organization size with attraction mode enabled. This mode makes the organizations much more separated and stable.

With the "escape" mode enabled, the entities move faster, and the changes in the PM stat are also rapid. In this case, the oscillations are greater, as shown in Fig. 8.6.

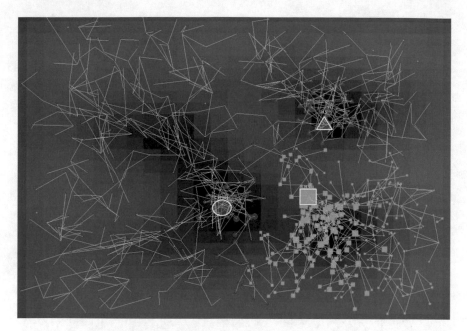

**Fig. 8.4** The organization image, attraction mode ON

## 8.5 Conclusion

The main conclusion is that no steady state is reached by the model and that the organizations are in permanent movement. This movement, after sufficient simulation time, is oscillatory, like the stable cycles in non-linear, orbitally stable dynamic systems (Chen [8], Weinstein [42]). The active components of this model are the individuals called entities (the monitor process is only an auxiliary component). The entities are "alive," executing their events. Though the decisions they take are very simple (where to appear on the political map, climb, etc.), they can be considered as agents of an agent-oriented simulation. Both object- and agent-oriented models provide interesting qualitative results that can be used as hints while dealing with the reality. As mentioned in the introduction, the historical data from the real world are similar to those obtained from our simulations.

The important advantage of such simulations is the possibility of obtaining results that can hardly be reached by other (analytical, sociological) methods. For example, how can be seen from the model description without simulating that the organization size will oscillate with a period of about 250-time units (Fig. 8.3)? Another advantage of the tool used here (Bluesss) is the open nature of the model. New events can be easily added to the entity process, reflecting a possible entity behavior and resulting in

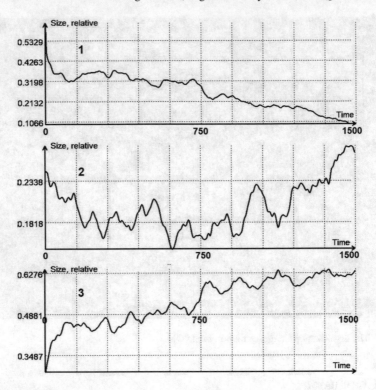

**Fig. 8.5** Relative size of the organizations, attraction mode ON

other, sometimes unexpected behavior of the organizations. This may be the topic of further research. Such research may be a statistical analysis of the simulation results and the series of simulation experiments with some kind of sensitivity analysis with respect to the properties of the entities defined in the model. This, however, should be done after defining a multi-dimensional and more realistic *political map* and greater, may be variable, number of created organizations with more flexible, variable structure. Some relation to terrorists and other illegal kinds of organizations may be pointed out by combining this model with, for example, that presented in Raczynski [31].

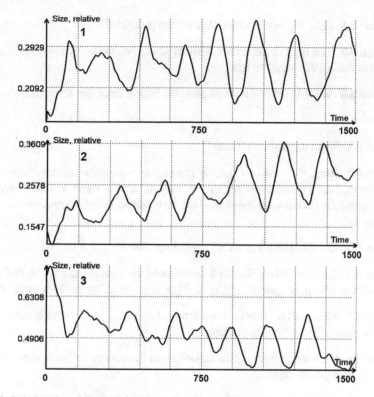

**Fig. 8.6**  Relative size of the organizations, escape mode ON

## 8.6   Questions and Answers

**Question 8.1**  The model discussed in this chapter can be characterized as: (select)
1. System Dynamics model
2. A model with distributed parameters
3. An agent-based model
4. A celular-automata model

**Question 8.2**  What is the *political map* (PM) in the model presented in this chapter?

**Question 8.3**  What modeling/simulation tool is used here?

**Question 8.4**  What is the *corruption field*?

**Question 8.5**  What is the *corruption level* of an agent?

**Question 8.6** How the organizational structure is defined?

**Question 8.7** The organizations are defined at the beginning of the simulation, or they grow during the program run?

**Question 8.8** What is the main result and conclusion from the simulations?

**Answers**

**Answer 8.1** 3. An agent-based model

**Answer 8.2** Here, the *political map* is the region where the agents move. It can be interpreted as a geographic region, as well as certain more abstract "space of ideas," where the coordinates the degree of, for example *religious ortodox*, *socialist*, *ecologist* etc.

**Answer 8.3** The BLUESSS simulation package, see Sect. 8.3

**Answer 8.4** The *corruption field* is a function of the coordinates at the PM (agent position) that tells how "good" or "spoiled" or "corrupted" is the spot in the PM.

**Answer 8.5** The agent *corruption level* is one of the agent's attributes. It takes values between "honest" and "totally corrupted."

**Answer 8.6** Each agent has several *subordinates* (pointers to other agents), and the pointer to its *superior agent*.

**Answer 8.7** No organization exists at the very beginning. The organizations grow during the simulation, as a self-organizing process. This is the result of the actions (micro-behavior) of the agents.

**Answer 8.8** The main result is that the organizations *almost never* reach any steady state or equilibrium. The agents constantly move over the PM, and the organization size enters in a well-defined oscillations. The influence of agent parameters and behavior patterns on the model macro-behavior is discussed.

# References

1. Adamic L, Adar E (2005) How to search a social network. Soc Netw 27(3):187–203
2. Anand V, Ashforth BE, Joshi A, Joshi M (2005) Business as usual: the acceptance and perpetuation of corruption in organizations. Acad Manag Exec 19(4)
3. Axelrod R (1997) The complexity of cooperation: agent-based models of competition and collaboration. Princeton University Press
4. Bak P (1997) How nature works: the science of self-organized criticality. University Press, Oxford
5. Borshchev A, Filippov A (2004) From system dynamics and discrete event to practical agent based modeling: reasons, techniques, tools. In: The 22nd international conference of the system dynamics society, July 25–29, 2004, Oxford, England

6. Caulkins PJ, Feichtinger G, Grass D, Hartl RF, Kort PM, Novak AJ, Seidl A (2013) Leading bureaucracies to the tipping point: an alternative model of multiple stable equilibrium levels of corruption. Eur J Oper Res 225:541–546
7. Chatterjee S, Seneta E (1977) Towards consensus: some convergence theorems on repeated averaging. J Appl Probab 14:89–97
8. Chen G (2004) Stability of nonlinear systems. Encyclopedia of RF and microwave engineering. Wiley, New York, pp 4881–4896
9. Cecconi F, Parisi D (1998) Individual versus social survival strategies. J Artif Soc Soc Simul 1(2)
10. Cioffi-Revilla C (1998) Politics and uncertainty: theory, models and applications. Cambridge University Press, Cambridge UK
11. Cohen J, Kejnal J, Newman C (1986) Approaching consensus can be delicate when positions harden. Stoch Process Appl 22:315–322
12. Daft RL (2013) Organization theory and design. South Western Cengage Learning. Erin Joyner, ISBN-13:978-1-111-22129-4
13. Crowder RM, Robinson MA, Hughes HPN, Sim YW (2012) The development of an agent-based modeling framework for simulating engineering team work. IEEE Trans Syst Man Cybern Part A Syst Hum 42(6):1425–1439
14. Deffuant G, Amblard F, Weisbuch G, Faure T (2002) How can extremism prevail? A study based on the relative agreement interaction model. J Artif Soc Soc Simul 5(4)
15. Edmonds B, Hernández C, Trotzsch K (eds) (2007) Social simulation: technologies, advances and new discoveries. Inf Sci Ref, Hershey, PA. ISBN 9781599045221 (pb)
16. Edmonds B (2012) Context in social simulation: why it can't be wished away. Comput Math Organ Theory 18(1):5–21
17. Elsenbroich C (2012) Explanation in agent-based modelling: functions, causality or mechanisms? J Artif Soc Soc Simul 15(3):1
18. Epstein JM, Axtell R (1996) Growing artificial societies: social science from the bottom up. Brookings Institution Press, Washington D.C
19. Galam S, Wonczak S (2000) Dictatorship from majority rule voting. Eur Phys J B 18:183–186
20. Gotts NM, Polhill JG, Law ANR (2003) Agent-based simulation in the study of social dilemmas. Artif Intell Rev 19(1):3–92
21. Holland JH (1998) Emergence: from chaos to order. Helix Books: Addison-Wesley Publishing Company, Massachusetts
22. Hughes HPN, Clegg CW, Robinson MA, Crowder RM (2012) Agent-based modelling and simulation: the potential contribution to organizational psychology. J Occup Organ Psychol 85:487–502. https://doi.org/10.1111/j.2044-8325.2012.02053.x
23. Krause U (2000) A discrete nonlinear and non-autonomous model of consensus formation. In: Elaydi S, Ladas G, Popenda J, Rakowski J (eds) Communications in difference equations, Amsterdam, Gordon and Breach, pp 227–236
24. Lambsdorff JG (2012) New advances in experimental research on corruption research in experimental economics. Emerald Group Publishing Limited, vol 15, pp 279–299. ISSN: 0193-2306, https://doi.org/10.1108/S0193-2306(2012)0000015012
25. Latane B, Nowak A (1997) Self-organizing social systems: necessary and sufficient conditions for the emergence of clustering, consolidation and continuing diversity. In: Barnett FJ, Boster FJ (eds) Progress in communication sciences. Ablex Publishing Corporation, pp 1–24
26. Long JE (2002) Systems analysis: a tool to understand and predict terrorist activities, Vitech Corporation, Web page link: http://www.seecforum.unisa.edu.au/Sete2002/ProceedingsDocs/62S-Long-INTEL.pdf
27. Lustick S (2000) Agent-based modeling of collective identity. J Artif Soc Soc Simul 3(1). http://jasss.soc.surrey.ac.uk/3/1/1.html
28. Macy MW, Willer W (2002) From factors to actors: computational sociology and agent-based modeling. Annu Rev Sociol 28:143–166

29. Moss de Oliveira S, Stauffer D (1999) Evolution, money, war and computers - non- traditional applications of computational statistical physics. Teubner, Stuttgart-Leipzig. PewResearch Center for the People & The Press, A closer look at the Parties in 2012, available from http://www.people-press.org/2012/08/23/a-closer-look-at-the-parties-in-2012/
30. Pinto J, Leana CC, Pil FK (2008) Corrupt organizations or organizations of corrupt individuals? two types of organization-level corruption. Acad Manag Rev 33(3):685–709
31. Raczynski S (2004) Simulation of the dynamic interactions between terror and anti-terror organizational structures. J Artif Soc Soc Simul 7(2). England. http://jasss.soc.surrey.ac.uk/7/2/8.html
32. Raczynski S (2006) A self-destruction game. J Nonlinear Dyn Psychol Life Sci 471–483
33. Raczynski S (2009) Discrete event approach to the classical system dynamics. In: Proceedings of the Huntsville simulation conference, the society for modeling and simulation, Alabama. Oct 28 & 29 2009, pp 254–258
34. Saam NJ, Harrer A (1999) Simulating norms, social inequality, and functional change in artificial societies. J Artif Soc Soc Simul 2(1)
35. Schachter S, Singer JE (1962) Cognitive, social, and physiological determinants of emotional state. Psychol Rev 69(5):379–399
36. Shelling T (1978) Micromotives and macrobehavior. W.W. Norton & Company, chapter 4, pp 137–166
37. Sibertin-Blanc C, Roggero P, Adreit F, Baldet B, Chapron P, El-Gemayel J, Mailliard M, Sandri S (2013) SocLab: a framework for the modeling, simulation and analysis of power in social organizations. J Artif Soc Soc Simul 16(4):8
38. Silverman E, Bijak J, Hilton J, Cao VD, Noble J (2013) When demography met social simulation: a tale of two modelling approaches. J Artif Soc Soc Simul 16(4):9
39. Staller A, Petta P (2001) Introducing emotions into the computational study of social norms: a first evaluation. J Art Soc Soc Simul 4(1)
40. Stocker R, Cornforth D, Bossomaier RJ (2002) Network structures and agreement in social network simulations. J Artif Soc Soc Simul 5(4)
41. Younger SM (2003) Discrete agent simulations of the effect of simple social structures on the benefits of resource sharing. J Artif Soc Soc Simul 6(3)
42. Weinstein MI (1986) Lyapunov stability of ground states of nonlinear dispersive evolution equations. Comm Pure Appl Math 39:51–67. https://doi.org/10.1002/cpa.3160390103

# Chapter 9
# The Space of Models, Semi-Discrete Events with Fuzzy Logic

## 9.1 Introduction

This chapter contains a report about some proposals and new insight on modeling and discrete event simulation. Thus, it is not written as a textbook, where the basic known facts are explained. Thus, the section "Questions and Answers" of this chapter is reduced, compared to that of the first chapters of the book.

To compare models, we should know how far is one model from another. So, we must define the distance between models. Such distance induces a topology in the space of models. Once we have a topology, we can verify if a sequence of models converges, when a model parameter approaches a given limit, and verify if the limit exists at all. This may also be useful in validity verification for some "idealized" models and may help to avoid situations when the limit is tied to a singularity in the model space.

Our distance is based on the Hausdorff set-to-set distance. This way we can compare models, analyze convergent sequences of models and handle the mappings from parameter space to the model space. The continuity of such mappings can be investigated. This is useful when selecting a simplified model specification, deciding if a model component can be removed or the model structure simplified.

Let recall the definition of the Hausdorff distance between sets.

Consider two non-empty subsets $X$ and $Y$ of a metric space. The *Hausdorff distance* between $X$ and $Y$ is defined as follows:

$$d_H(X, Y) = max \left\{ \sup_{x \in X} \inf_{y \in Y} d(x, y), \sup_{y \in Y} \inf_{x \in X} d(x, y) \right\}, \qquad (9.1)$$

where $sup$ represents the least upper bound, and $d(*,*)$ is the distance between two points. The Hausdorff distance permits to use the concept of continuity of set-valued functions. We say that a mapping from the real line to the space of closed subsets of $R^n$ is *continuous in Hausdorff sense* if it is continuous in the sense of the Hausdorff distance (in the topology induced by the Hausdorff distance).

© The Author(s), under exclusive license to Springer Nature Switzerland AG 2022
S. Raczynski, *Models for Research and Understanding*, Simulation Foundations, Methods and Applications, https://doi.org/10.1007/978-3-031-11926-2_9

If we do not impose some additional conditions, $d$ is not a distance, but rather a semi-distance in the given set of sets. A simple example is the distance between an open set $U$ and its closure $W$. According to the definition, $d(U, W) = 0$, but the two sets are not equal to each other.

Another definition that we will need is a distance between two functions of time. Consider the set $G$ of all integrable functions $[0, T] \rightarrow R$, where $[0, T]$ is a closed interval of real numbers between 0 and $T$. The distance we will use is defined as follows:

$$h(f, g) = \int_0^t (f(y) - g(y))^2 dy, \qquad (9.2)$$

where $f, g \in G$.

## 9.1.1  Distance Between Models

There are many ways to compare models. One could say that two models are similar to each other if they have similar structures, use similar distributions of random numbers or reveal similar behavior of the state variables or of the output variables. Since we want to compare models of different structures or models with variable structures, the model structure can hardly be used as a factor in the definition of the distance. Two models with very similar structures can behave in a very different way, and even small changes in the random variable generator inside the model can imply big changes in simulation results. Another way to define the distance is to treat models as "black boxes" and suppose that the observer can only measure the initial conditions, and input and output variables. We use this approach in the proposed distance definition.

Here, we will use the simplest, though limited concept of the distance between models. Consider two models that start from the same initial conditions, run over the same model time interval, and produce results in the form of one or more functions of time. The distance can be calculated with the following assumptions:

Case 1. Both models provide N plots of functions of one variable, in response to the same input signals. Then, the distance between models is calculated as the sum of distances between corresponding functions, calculated according to (9.2).

Case 2. If the models produce a series of functions of multiple variables, calculate the Hausdorff distance between the multi-dimensional surfaces (sets of points) given by the plots.

Case 3. The models are stochastic, and produce plots like in case one and two, different in each run. In this case, calculate the corresponding probability density function(s) and calculate the distance in the similar way as in case one and two.

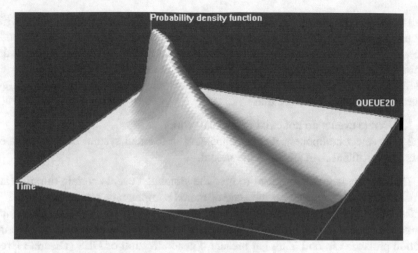

**Fig. 9.1** The probability density function of the length of a queue

Figure 9.1 shows an example of a density function that can be used to model comparison. This is the probability density function for the length of a waiting line. Vertical axis is the density value, as function of the queue length and the time.

## 9.2   Strictly Discrete Event Model

Strictly discrete events. In the *strictly discrete event model*, the actions of model components occur in some time instants (discrete events). In Chap. 7 we discuss such kinds of models. Let's recall here some basic concepts.

The commonly used discrete event simulation like Arena, Promodel, or Simul8, uses the strictly discrete event model. A theoretical base for such kind of models has been formulated in the Discrete Event Specification Formalism (DEVS), see Sect. 7.4. Using DEVS, one can construct complex models from simple "atomic" models, and couple such models in a hierarchical way. The definition and applications of DEVS may be found in many sources, see, for example, Barros [1], Chow [2], Zeigler [15], see Chap. 7. The basic DEVS concept is the model specification. The specification elements are the model input space, the state space, output space, internal and external state transition functions, and the output function that defines the model output as a function of the total model state. This defines an elemental "atomic" model. Model coupling is an important topic in DEVS theory, which provides the basic theoretical frame for complex model creation.

Both the simple discrete event simulation, widely used from the early 1960s, and the DEVS formalism, represent the common platform for many applications and lead to fast implementations. However, in this approach some problems arise. While coupling small "atomic" models, a new element called *select* must be added to handle the execution of simultaneous events, if any. Select resolves possible conflicts and ambiguity in simultaneous event execution, see Saadawi [11]. The necessity for the select element needs special treatment that is the topic of serious research work, see sources cited above. Observe that

1. Discrete events do not exist in the real world.
2. The select component has no equivalent in the real system being simulated. This is an artificial item added to the model.

This chapter is just a proposal to give the simulated events certain duration and to use fuzzy logic to represent the state of model components, see Zadeh [14].

Recently, several papers on fuzzy discrete event simulation appear. Perrone and Zinno [9] discuss the problem of processing fuzzy data within a discrete event simulation process. Lin and Ying [6] present a generalization of DES (Discrete Event Simulation) to fuzzy event systems. The model observability is discussed and some medical applications pointed out. An application of fuzzy logic in uncertainties, vagueness and human subjective observation, and present fuzzy version of DES is also discussed.

Du and Ying [4] discuss vagueness and imprecision concerning states and event transitions of DESs where the membership grades are computed via fuzzy logic. An application to computerized human immunodeficiency simulation is presented. Zhang and Tam [16] use fuzzy logic to treat uncertain information input, vagueness, imprecision, and subjectivity in the estimation of activity duration, especially when insufficient or no sample data are available.

An application of discrete event simulation with fuzzy logic applied to production control is also discussed by Dassisti and Galatucci [3]. They propose a "pseudo-fuzzy discrete simulation," where the fuzziness is traced through a set of several classic simulations. The idea is to use the simulator as a fuzzy operator that includes some stochastic functions. Sadeghi et al. [12] propose to integrate the classic discrete event simulation with fuzzy logic. The authors point out that their approach may increase the accuracy of the simulation. This can be achieved by capturing imprecise, subjective, and linguistically expressed knowledge in the simulation inputs using fuzzy numbers.

Santucci and Capocchi [13] deal with an approach based on the use of Fuzzy Control Language allowing to facilitate the modeling and simulation of Discrete Event Systems involving uncertainty. The main contribution is to integrate the proposed language with the DEVS formalism.

The main objective of this chapter is to show an alternative approach to event simulation and to the treatment of simultaneous events. This is a methodological and conceptual framework rather than a proposal of a practical implementation. In the classical discrete event model, an event executes in a specific time instant, changing the state of the model. In our approach the event is a process that has a finite duration. Moreover, the event is something more than a change of the model state. Our event has, in addition, a goal to achieve (seize a server, enter a buffer, etc.). The goal achievement is a fuzzy variable that reaches the value of one when the goal has been reached. The event terminates its activity when the goal has been reached, or when its pre-defined duration period has elapsed and the goal variable is not growing.

## 9.3 Finite-Time Event Model

In the exact discrete event models, the events occur in discrete time instants. Each event executes a piece of code that may change its own state or the state of other components. Here, we assume that the model events are executed over a finite, non-zero time interval. The time interval of an event may be very small, resembling the discrete event models, or quite large. This type of event simulation will be called here *Finite-Time Event Specification* (FTES). The simulation that implements models with very short event duration (compared with the simulation final time) is called *semi-discrete event* (SDE) simulation. Consider a model with only one event, which starts at model time zero and terminates at the simulation final time. Such event may describe a continuous model defined by a set of differential equations. So, there is no conceptual difference between continuous and semi-discrete, SDE event model. However, we do not consider events with execution time equal to zero.

To define a simulation model, we must define its components, resources, and inter-action rules. Here, we consider models with an object- or agent-oriented simulation paradigm (Fishwick, [5]). Roughly speaking, the model components are active enti-ties that are created, and then go through the model resources executing their events. The entity is an instant of a generic entity class, where the entity events are declared. Each event activation occurs due to the execution of a corresponding event message. The simulation program manages the event message queue according to the pre-defined clock mechanism. This is nothing new, this paradigm has been established decades ago, see, for example, O'Keefe [8]. So, events are actions of the model entities. The event message contains the event identification (the entity class, entity instant, event id and the event start time). Here, we do not enter in detail of a particular software implementation, rather treating with concepts of the FTES.

In addition to the event duration parameter (always non-zero), we require that every event represents an action that must have a well-defined aim or *event objective*. The event specification must include an algorithm that intends to achieve the aim of the event. The result of this action is measured by an event attribute that starts with value zero (not achieved) and reaches one if the aim is fulfilled. This attribute is interpreted as a fuzzy logic variable, which tells us the "degree" of the *aim fulfillment*. The objective may be achieved or not. So, another attribute of the event must be the abandon switch and the abandon function. If the abandon switch is set equal to 1 (true), then the event action stops, and the abandon function is executed.

### 9.3.1   The Chicken Game

In the game called *Chicken*, two people, say Jim and Fred, drive their cars toward each other from opposite ends of the road. If one of them swerves before the other, he is called a chicken, and the other wins. Of course, if neither swerves, they will crash. So, the outcomes may be: 1. Fred wins, 2. Jim wins, 3. The cars crash.

In the model used here, we do not pretend to exactly simulate what appears in the real game. This is just a simple example used to investigate the convergence of a sequence of models in the model space.

The model rules are the following:

Car attributes:

Position$(x, y)$, where $x$ is the distance, $y$ is the deviation from the initial straight line.

Velocities $v_x$ and $v_y$, in the direction $x$ and $y$, respectively.

Fear factor $f_g$. This is the fear growth rate. At certain fear threshold $f_t$, the player starts his decision event.

$D$—the duration of the "decision event," that is, the event of decision-making

The initial positions for Fred and Jim are $(-1, 0)$ and $(1, 0)$, respectively. Initial velocity of Fred is positive, the velocity of Jim is negative (they are going toward each other).

After the game starts, the fear level of the two players begins to grow, and their distance decreases. For each player, if his fear reaches the threshold level, he triggers his decision-making event. During this event, the driver observes the movement of the competitor. If the other player quits (his $y$ coordinate becomes different from zero), the driver follows with the straight movement. If the other player did not quit, the driver may take the decision to turn out. A "human factor" is taken into account. This means that the driver needs some (may be very small) time interval to decide if he quits or not. The decision is taken at some time instant inside the decision-making event. In addition, due to the human factor, the decision may be wrong or not be taken at all. This means that the driver may quit even if the other did, or he can follow the straight movement when the other does the same. Figure 9.2 shows the increasing fear and the decision-making event according to the car movement. The endpoint is where the possible crash may occur.

The decision event is an approximation of a continuous process. However, in our (recent) computers nothing is continuous. Thus, the event is executed in some small time steps, according to the following algorithm.

At each time-step, the player does the following:

1. If the other player quits, the player goes ahead. If the player velocity $v_y$ is positive, it can be updated to be zero and the car follows the straight line. This occurs with probability equal to 0.1 (in each step).

2. If the other player did not quit, the actual player quits with probability equal to $1/n$ at each time-step ($n$ is the number of time steps).

Note that the last decision has the probability $1/n$ in each step, which means that the decision can be taken at some moment inside the decision event. The total

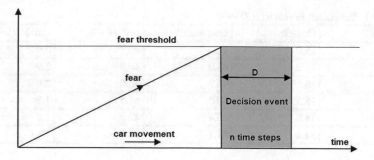

**Fig. 9.2** Car movement, increasing fear and the decisions event

probability of taking the decision is approximately equal to one. As for the point 1, there is some uncertainty about the decision. It may result to be mistaken, due to the human error.

If the two players have different fear factors or fear thresholds, the player that first enters the decision event will quit with high probability. This probability decreases if the decision events overlap. Here, we focus on the case when the two players have identical parameters. This means that the decision events start and terminate in the same time instants. With such, strictly overlapping events, the final outcome is uncertain.

Assume the following model parameters (equal for the two players):

Absolute initial velocity $v_x = 0.1$, $v_y = 0$

Fear grow rate $f_g = 0.1$

Fear threshold $f_t = 0.6$

Decision event duration $D = 0.5$

Number of steps in decision event $n = 50$

The simulation of the car movement is rather trivial and will not be discussed. What is relevant here is the decision event execution and final outcome. The simulation with the above parameters provides the following results:

Crash probability: 0.150

Fred wins: 0.424

Jim wins: 0.407

Crash due to false turn decision: 0.02

These are results obtained for the average the outcome from 5000 repetitions of the simulation run.

As for the reproducibility of the simulation, it should be mentioned that the simulation program has been coded in the BLUESSS simulation language see
http://www.raczynski.com/pn/bluesss.htm .

The simulation is rather simple. It consists in simulating the movement of each car in one or two directions, with uniform movement and detecting the crash.

**Table 9.1** Simulation results with $D \rightarrow 0$

| IntervalD | Crash% | Fred wins% | Jim wins% | Crash%(bad turn) |
|-----------|--------|------------|-----------|------------------|
| 1.0 | 15.0 | 42.4 | 40.7 | 2.0 |
| 0.5 | 16.2 | 41.0 | 40.9 | 1.9 |
| 0.1 | 16.1 | 42.3 | 40.0 | 2.0 |
| 0.01 | 15.5 | 40.9 | 41.5 | 1.8 |
| 0.001 | 15.8 | 40.4 | 42.5 | 2.0 |
| 0 | 34.1 | 23.1 | 22.0 | 19.7 |

Observe that looking at the rules of the game, we can see that the outcome does not depend on the decision event duration $D$. Indeed, running the same experiment with different values of $D$, we obtain the same results (up to small random fluctuations of the statistics). Now, suppose that $D = 0$. Taking this value, the model operations do not change. Simply all $n$ steps of the decision event are executed at the same model time instant. The time advance step inside the event becomes equal to zero, but there are still $n$ steps, so the algorithm cannot stop in a "zero-advance" loop.

Consider a sequence of models with $D$ approaching zero. As mentioned before, the "limit model" with $D = 0$ is the discrete event model with simultaneous events. It might appear that the final outcome for the limit case $D = 0$ should be the same. However, this is not the case. With $D = 0$ we obtain the following:

Crash probability: 0.341

Fred wins: 0.231

Jim wins: 0.220

Crash due to false turn decision: 0.197

The other results also reveal discontinuity at $D = 0$, as shown in Table 9.1.

Now, define the distance between models with different values of $D$. In this case, this is rather trivial; we take the difference in the model outcome, namely the average of crash occurrences. Using this distance, we can see that there is a singularity in the space of models (with different $D$), at the point $D = 0$.

Figure 9.3, as well as Table 9.1, show the simulation results with $D$ tending to zero, including the limit point $D = 0$. It can be clearly seen that the last point (discrete event model) represents a singularity in our model space.

### 9.3.2   Semi-Discrete Model Specification

Consider the following event specification. Here, a simulation that implements models with very short event duration (compared with the simulation final time) is called semi-discrete-event (SDE) simulation.

event = {ED, TG, AC, AV, AS, AA}

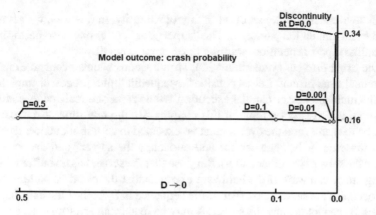

**Fig. 9.3**  Model outcome with D approaching zero

The attributes are as follows:

ED—event duration time, greater than zero

TG—event objective

AC—event action: a function or algorithm that is executed in order to achieve the objective

AV—a fuzzy logic variable with values between zero (no success) and one (the objective achieved)

AS—abandon switch. It is set to one ("ON" state) when the event objective cannot be reached.

AA—abandon action. This action is executed when the AS is ON.

Note that this is not an event message, so the event start time is not specified here. The above specification describes an event, and not the entity. Any FTES entity can have its own attributes, like the entity state XE, for example "on service (XE = 1)" or "out (XE = 0)." The state of the whole model is visible from the AC algorithm, so the actions may be modified according to the state of other entities. The FTES events resemble the "time events" defined in the classic discrete event simulation. Note that during the simulation entities interact with each other. In particular, an entity event can change the state of another entity and generate new event messages in the event queue. As for the fuzzy variables, they have been used in some discrete event models, see, for example, Lin and Ying [6].

Recall that the discrete event strategy also considers the *state events* that are invoked due to the changes in the model state. In FTES there is no need to introduce a different mechanism for the state events. To execute such events, we can create a "supervisor" entity object with only one event that starts at the simulation start time, and ends at the simulation final time. The component AC of this event receives the information about the model state, so it can trigger (put into the message queue with immediate event start time) an event of any other entity. In this way, we get a more unified event simulation paradigm. The "supervisor" entity may also perform other actions, like report generation, warning or error message display, etc.

In the exact discrete event simulation, simultaneous events occur in exactly the same model time instant. Let us consider a sequential, one-processor implementation. The multi-processor parallel execution would rather be realized as distributed simulation, which is not a topic of this chapter. So, the real time of the execution cannot be the same, and the events must be executed in certain sequence. For example, let the event A be "end service and unoccupy the server" and the event B be "occupy the server" (by other entity). Suppose that the server has no buffer where the entering entity can wait. If the hardware executes first the event A, and then B, the entity occupies the server. However, if the sequence is B-A, then the entering entity (of event B) cannot occupy the server; it may be lost, or an error may occur. This is something that does not happen in the real system. To manage such situations and other ambiguities in event execution, in the DEVS formalism the *select element* has been introduced.

Now, consider the corresponding FTES model. Let A and B be simultaneous events with identical start time and duration interval. The aim of the event B is to occupy the server, and that of event A is unoccupy it. In FTES we have no ambiguity. When the event B starts, its occupation fuzzy variable AV starts to grow. Simultaneously, the AV variable of event A decreases. Finally, AV of A reaches zero, and AV of B reaches 1. The events terminate with well-defined final model state.

There may be other situations where the final state is ambiguous, even in FTES simulation. If, for example, two identical simultaneous events intent to occupy the same server, then the result may depend on the implementation. Observe, however, that this ambiguity is originated in the real system, and it is not the result of the model event execution. In other words, the *ambiguity of the real system cannot be eliminated*, and should not be eliminated by any artificial model component defined by the user.

The FTES entities run simultaneously in the model time. So, it is difficult to define a unique program flow diagram or algorithm that treats simultaneous events. To see how this may work, we must define some event execution rules. A possible (and not unique) rule definition may be as follows:

1. When the event starts, its achievement variable AV is set equal to zero, the abandon switch is set to zero, and the AC algorithm is invoked. AC runs and changes the value of aim achievement variable AV, due to the objective fulfillment. It also may perform other actions.

2. When the event execution time reaches its duration time then:

(a) if the value of AV does not grow or reaches one, the event terminates.

(b) if the value of AV is growing, then the event execution continues, until AV reaches one or stops growing.

3. The event execution terminates when one of the following conditions occur.

(a) the AV variable reaches one.

(b) the value of AV is less than one, it does not grow and the event execution time exceeds the event duration.

(c) the AS switch is set to 1.

4. If the event terminates for any reason and the value of AV is less than one, then the abandon action AC is invoked.

Consider a simple event, which goal is to occupy a server (a model resource). The AC algorithm may be defined by the following function (SO is a fuzzy logic variable that represents the server occupation). AV variable is the actual server occupation by the current model entity (a client that intends to occupy the server).

$AV = (t - t_0)/$ ED when AV and SO are less than one. ED is the event duration.

AV stops to grow if SO reaches one (i.e., when the server becomes completely occupied by another entity)

In the above formula, $t$ is the model time and $t_o$ is the event starting time. If two entities start to occupy the same server, then the server occupation SO is given by the fuzzy OR operator of the two AV variables. In the case when only one entity intents to occupy the server, we have SO = AV.

Other situations similar to the DES simultaneous events may occur when the events overlap but are not exactly identical. In these cases, FTES does not generate any ambiguity. The examples of the semi-discrete overlapping events are given further on.

## 9.3.3  Model Coupling

Model coupling is not the main topic of this chapter. However, let us make some remarks on this. A FTES model consists of one or more entity class declarations. Using the object-oriented programming terminology (like C++ or Delphi), each class has a set of public and private variables (attributes), and one or more methods. Due to the class declaration, the corresponding objects are generated and activated at the runtime, being the FTES model entities. The methods represent the events that may occur in the entity "life" inside the model.

The clock mechanism and event queue management is a pre-defined task of the simulation system, not of the user model definition. The "supervisor" class is just one more class of the model, with only one event method and only one instant created at the runtime. Note that a FTES event is not executed in one discrete time instant, but it must run over certain model time interval, like a continuous process. This can be done, but the necessary code will be a little bit more complicated compared with the classic DES simulation.

Now, consider two or more FTES models. To couple the models, we can create a new program that simply includes all class definitions of the original models (the sub-models). Doing this, we obtain a model with two or more supervisor classes. As these classes have the same structure as any other model class, we can treat them as a "regular" model entity specifications. This means that we need a new "supervisor" entity. This entity will activate the supervisor entities of the sub-models. Now, supervisor entities of the sub-models may start at any time instant, different from zero, with duration ED different from the whole simulation period. Moreover, these entities can be activated several times, and two or more sub-models of the same type may run concurrently. Repeating this procedure, we can create a hierarchy of models, including the previously defined ones.

In such way, FTES provides a unified frame for both semi-discrete events and continuous simulation, which can run concurrently as the sub-models of the same simulation program.

## 9.4  More Examples

It should be noted that the model proposed here is a conceptual and theoretic one, rather than a ready-to-implement simulation scheme. The use of semi-discrete events may perhaps eliminate the problem of simultaneous events of DEVS. The term Semi-Discrete Event (SDE) suggests that the event duration is small compared to the total simulation period. In fact the event can have any duration. Here, we focus on SDEs because the examples and most of the other issues in this chapter refer to models normally treated with the discrete event (of zero duration) simulation.

As for the state of the model component and the rules of interaction, we use the fuzzy set logic (Zadeh [14], Novak [7]).

The SDE event message in a possible implementation should contain at least the information about the event identification and its start time. The aim of the event is an important item, given in the event description. The event can terminate if one of the following conditions is fulfilled.

1. The aim variable is equal to one.

2. The aim variable does not increase at the moment and the event duration expires.

3. The abandon switch $AS$ is set equal to 1, possibly as a result of the action of other entity.

In the examples below, the objective of the events TG (see Sect. 9.3.2) is defined for each event, e.g., "seize the server," "release the server" or "exist." The latter target refers to entities that enter the model or disappear. The event action AC is the process of seizing the serve, entering the model, etc. These actions result in changes of the corresponding target achievement variable AV, supposed to have a linearly increasing or decreasing value.

### 9.4.1  Example 1: One Server

Suppose that, according to our semi-discrete event rules, an arriving entity that is seizing a server will do it until it succeeds, or until the arrive event expires without improvement of the aim variable. Below, the entity X is arriving and the entity Y is the entity actually on the server, leaving it.

The state of the entity contains a fuzzy logic variable, with values inside the interval [0, 1]. It may be "on service" ($XE = 1$) or "out" ($XE = 0$). It also can be equal to, say, 0.7 when the state is uncertain, most likely to be "in service." Moreover, as will be shown later on, the entity may occupy one or more servers at a time, with occupation values being, for example 0.4 and 0.6, respectively (two server cases), defined by two state components, $XE1$ and $XE2$.

The SDE events may be separated in time from each other, or overlapped. We will call SDEs being strictly simultaneous if they have the same start time and duration. The two or more SDEs will be called simultaneous if their execution periods overlap. Let us use the fuzzy logic and permit that the server changes from "busy" to "free" continuously, passing through all intermediate state values. The state parameter of the server is given by the function *free(time)*. This may be any monotonic function of time, changing from 0 to 1. The entity observes the server, and occupies it due to the function seize(time) being, in this case, equal to $1 - free(time)$. The interaction rules of the model are as follows. The entity arrives at $time = TS1$ and starts observing the server. If after $TS1 + T1$ the entity state is "out," the entity terminates its activity. However, if between $TS1$ and $TS1 + T1$ the server begins to change its state to "free" ($free(time) > 0.0$). The entity occupies the server to the same degree as it becomes free. The entity remains active as long as it occupies the server (with fuzzy occupation $> 0.0$), until it reaches the state "on service" ($XE = 1$).

Let the two events are strictly simultaneous. No ambiguity exists in this case, see Fig. 9.4. The arriving one is Entity X, and that which releases the server is Entity Y. The horizontal axis is the time, and the shadowed areas show the fuzzy logic values between 0 and 1. In accordance with the logic of the real system, no ambiguity or strange behavior occurs. The arriving entity observes the server and occupies it "continuously" while the other entity leaves it. The degree of occupation is a fuzzy variable, changing from 0 to 1. Those are simultaneous events, but not strictly simultaneous. When the entity arrives, the server is still occupied. Just before the arrive event terminates, entity Y starts leaving the server. Entity X seizes the server (partially), so it does not deactivate when its arrive event terminates.

One can expect that if $T1$ approaches zero then, in the limit, the model becomes a discrete event one when the event's duration approaches to zero. This, however, is not the case. We have a discontinuity at $T1 = 0$, because the outcome of the discrete version is 50% lost entities, or undefined result. There is a singularity at the point $T1 = 0$. My point is that in the general space of models (Raczynski, [10]) the discrete event models represent a singularity. Now, consider a model with two entities that enter the system.

Let us define the following fuzzy variables (all of them in [0.1] ):

$XEX$—the "existence" of the entity $X$.

$YOS$—entity $Y$ on server

$XOS$—entity $X$ on server (this is the aim variable)

$SFR$—server free

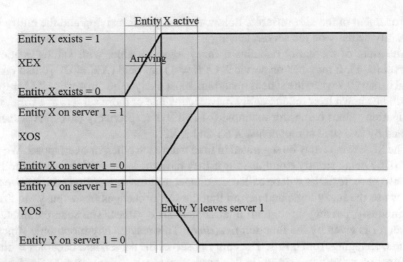

**Fig. 9.4** Simultaneous events. One server

Suppose that the entity $X$ arrives at time instant $T_{X1}$, and that the duration of the "arrive" event is equal to $TX$. The entity $Y$ starts to leave the server at $t = T_{Y1}$ and takes the time interval $TY$ to release it completely. $XEX(t)$ may be defined as any monotonic function of time, changing from 0 to 1. Let us define it as follows:

$$XEX(t) = \begin{cases} 0 \text{ for } t \leq T \\ \dfrac{t - T_{X1}}{T_X} \text{ for } T_{X1} < t < T_{X1} + T_X \\ 1 \text{ for } t \geq T_{X1} + T_X \end{cases} \qquad (9.3)$$

In the similar way, we define the function $YOS(t)$, changing from 1 to 0 in the interval $[TY1, TY1 + TY]$.

Once the entity appears ($XEX > 0$), it starts to observe the server. It occupies it (in some degree) if $YOS$ is less than 1. The entity remains active until it has occupied the server completely ($XOS = 1$). The variable $XOS$ cannot be greater than $XEX$, and cannot be greater than $1 - YOS$. So, $XOS = XEX$ AND $(1 - YOS)$, where AND is the fuzzy logic "and" operator. Figure 9.4 shows an example of simultaneous (but not strictly simultaneous) events. Entity X is arriving within the interval marked as "arriving." When it appears, the server is still fully occupied by the entity Y. Just before the end of the arriving interval the entity Y is beginning to leave the server, so the entity X starts to occupy it. The activities of entity X terminate when it has fully seized the server. The process of Fig. 9.5 is different; now, the entity is arriving slowly. The variable $XOS$ starts to grow fast, but after it reaches the level $XEX$, the growth is slower because its presence on the server cannot be greater than its presence $XEX$ (fuzzy AND, as above).

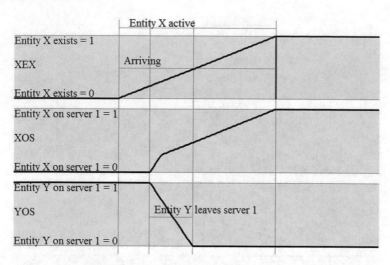

**Fig. 9.5** Simultaneous events. One server, slow arrival

## 9.4.2 Example 2: Two Servers

In Fig. 9.6 we see a similar process with two servers that are released simultaneously. These are simultaneous, but not strictly simultaneous events. At the beginning, server 1 is occupied by entity Y and server 2 by entity Z. Just before the entity has arrived, entity Z starts to release server 2. Entity Y starts to release server 1 after the entity X has arrived. Server 1 is being released faster than server 2. So, entity X first occupies (partially) server 2. When server 1 is being released, entity X occupies also server 1. The entity Y leaves server 1 quickly, so the entity X fully occupies it and terminates its activity.

We have the following fuzzy variables:

$XEX$—entity X present
$YOS$—entity Y on server 1
$ZOS$—entity Z on server 2
$XOS$—entity X on server 1
$XOS2$—entity X on server 2

The entity remains active until one of the fuzzy variables $XOS1$ or $XOS2$ reaches value 1. We use fuzzy logic, so the entity can (partially) occupy both servers at the time.

Entity X arrives at time instant $TX_1$, and that the duration of the "arrive" event id equal to $TX$. Entity Y starts to leave server 1 at $t = TY_1$ and takes the time interval $TY$ to release it completely. Entity Z starts to leave server 2 at $t = TZ_1$ and takes the time interval $TZ$ to release it completely. Entity X observes the two servers The variables must obey the following relations:

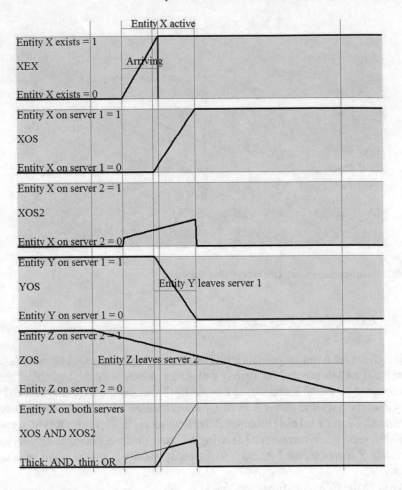

**Fig. 9.6** Simultaneous events. Two servers

$XEX(t)$—given by the formula (9.3)

$$YOS(t) = \begin{cases} 1 \text{ for } t \le T_{Y1} \\ \dfrac{T_{Y1} - t + T_Y}{T_Y} \text{ for } T_{Y1} + T_Y \\ 0 \text{ for } \ge T_{Y1} + T_Y \end{cases} \qquad (9.4)$$

$$ZOS(t) = \begin{cases} 1 \text{ for } t \le T_{Z1} \\ \dfrac{T_{Z1} - t + T_Z}{T_Z} \text{ for } T_{Z1} + T_Z \\ 0 \text{ for } \ge T_{Z1} + T_Z \end{cases} \qquad (9.5)$$

$XOS = XEX$ AND (NOT $YOS$)

$XOS2 = XEX$ AND (NOT $ZOS$)

Note that the server occupation $XOS$ and $XOS2$ cannot be greater than the presence variable $XEX$ of entity X.

Here "AND" and "NOT" are fuzzy logic operators. Entity X may occupy the two servers simultaneously (up to some degree). The expression ($XOS$ and $XOS2$) indicates such a simultaneous occupation.

### 9.4.3 Example 3: A Battlefield

Consider a situation, when there are four entities, two of them friendly: the tank F1 and the aircraft F2, and two enemies: a tank E1 and an aircraft E2 (see Fig. 9.7). Four discrete events are scheduled to be executed at the same time instant: F1 shoots and eliminates E1, E1 shoots and eliminates F2, F2 shoots and eliminates E2, E2 shoots and eliminates F1. The aim of the shooting event is to hit. Obviously, an entity that has been eliminated cannot execute its shooting event. If we assign any priorities to the events or if the event execution is random or hardware-dependent, some of the entities will survive. However, in the real system, if the entities shoot simultaneously, then all of them should be eliminated. A simple solution in discrete event simulation is to divide each event in two: "shoot," and "being hit" (the event of the target), the

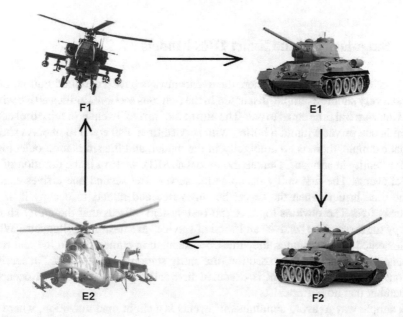

**Fig. 9.7** A battlefield

last one executed slightly after the first one. However, this will duplicate the number of events.

Now, consider a similar model with semi-discrete events. The four entities execute their "shoot" events simultaneously. There are eight fuzzy logical variables:

$FnS$—entity Fn shoots ($n = 1, 2$). The variable changes from 0 to 1, where 1 means hit.

$EnS$—entity En shoots

$FnE$—entity Fn exists. If the entity is shooting then FnE changes from 1 to 0, where 0 means "destroyed."

$EnE$—entity En exists

Obviously, an entity that has been destroyed does not exist and cannot shoot. As in the previous model, it is supposed that once initiated an event, it cannot be interrupted. The other rules are simple: The fuzzy logical variables change from 0 to 1 or from 1 to 0 due to the corresponding monotonic functions of time, like that of the Eq. (9.3). Running this model with strictly simultaneous events, the final score is zero survivals, as expected. No ambiguity arises. Figure 9.8 shows the plots of the model variables for simultaneous, but not strictly simultaneous events (when all events overlap, with slightly different start points and event duration. Whatever would be the placement of the event intervals, the final score is the same: no survivors. Again, note that the discrete event version of the model with event duration equal to zero is not the limit case of the semi-discrete model with event duration approaching zero. In the crisp discrete version, there are survivals, and in the SDE we have zero survivals, independently of the event duration (supposing the events overlap).

## 9.5   Singularity of the Exact DES Models

Let us take a look at a model where the events always have a certain duration. Let's repeat a very simple example, discussed in the previous section. Entities arrive with a fixed inter-arrival time equal to one. The aim of the "arrive" event is to seize the server. There is one server without a buffer, with service time also equal to one. As shown in that example, there is no ambiguity in the model, and the simulation outcome is always "entity in service." Denote the model as M(I), where I is the duration of the model events. The first entity occupies the server. The second one arrives exactly in the time instant when the server becomes free and intents to occupy it, if the server is free. The obvious logic of the real system tells us that the entity should occupy the server. The "arrive" and "end of service" events are simultaneous. While implemented on a computer, they must be executed in some order in the real time. Depending on the order of execution, the entity state may result to be "in service" or "out" if the "arrive" event is executed first and finds the server still occupied (remember that no buffer exists).

A simple way to avoid simultaneous events is a slight randomization, where the events are executed in the exact time instant, plus a small random value. However, in this case approximately 50% of entities will become "out" (cannot occupy the

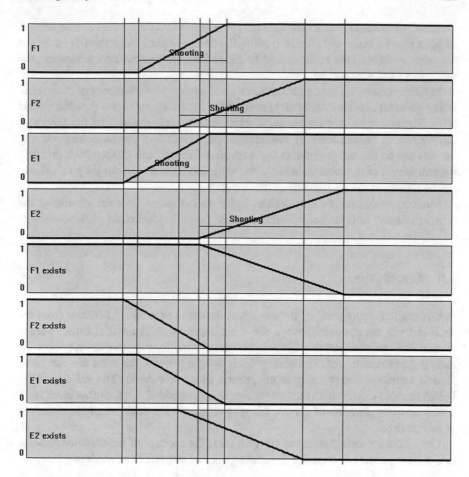

**Fig. 9.8** A battlefield. Simultaneous (overlapping) events

server). If we let the hardware or the operating system decide which execution order will be applied, then the result is undefined at all. Anyway, if the model is supposed to be valid, the result cannot depend on the particular implementation. As the consequence, we must add the selection rule to the model, as is done in the DEVS model specification, see Zeigler [15] and Chow [2]. This may be avoided if we admit that the event is executed within some finite time interval, rather than in a discrete time instant with zero duration.

Let R(I) be the result of the simulation run.

As stated above, the outcome R(0) is "50% out" or undefined, implementation-dependent result. Now, consider a sequence of models M(I) with I approaching zero. For any I > 0 the outcome R(I) is "100% in service," but in the limit point I = 0 the result is completely different. So, in the sense of the simulation result R(I), the discrete model (I = 0) is not the limit case of the sequence of models with I > 0. In other words, we have a discontinuity in $I = 0$. This may have important theoretical consequences. In other words, while M(I) tends to M(0), the corresponding outcome R(I) does not converge to R(0). This issue should be taken into account while considering the validity of the DES models.

In further research, the topic should be treated by means of the theory of categories, which gives an excellent theoretical support for modeling problems (see Raczynski [10]). The concepts of morphisms, functors, and transformations of the theory of categories can be comfortably interpreted and included in the modeling process. Models can be treated as functors that convert the morphisms of the real world into the morphisms of the models. While treating the model as a new category, its validity can be related to the correctness of the category.

More examples on the singularities in the model space, like the models of the "chicken game" or three.body collision can be found in Raczynski [10].

## 9.6   Conclusion

In this chapter, a proposal of the event simulation is presented, different from the strict discrete event models, when the event duration, in the model time, is equal to zero. It is pointed out that the appearance of simultaneous events may provoke ambiguities in the model. The main point is that, in general, the strict discrete event models represent a singularity in the general space of models. The validity of the DEVS coupled models with the *select* element is questioned. On the other hand, if we admit the finite duration of events, this may eliminate the problem of simultaneous event handling.

One of the important tenets of this chapter is the concept of the *distance between models* that permits to consider the convergence of sequences of models.

## 9.7   Questions and Answers

**Question 9.1** How the distance between sets can be calculated?

**Question 9.2** How we define the distance between models?

**Question 9.3** What is the *finite-time event model*?

**Question 9.4** Why the discrete event models represent a singularity in the space of dynamic models?

**Answers**

**Answer 9.1** We use the Hausdorff set-to-set distance, see Eq. (9.1).

**Answer 9.2** We calculate the distance between the corresponding sets of output data, using the Hausdorff set-to-set distance.

**Answer 9.3** We assume that the model events are executed over a finite, non-zero time interval. The time interval of an event may be very small, resembling the discrete event models, or quite large. This type of event simulation is called here *Finite-Time Event Specification* (FTES).

**Answer 9.4** Let $P$ be one of the parameters of model $M(P)$. Consider the sequence $\{M(P_k), \ k \to \infty\}$ and suppose that the sequence converges. Let $M(0)$ produces the result $R^0$ and $\{M(P_k), \ k \to \infty\}$ produces the sequence of results $\{R_k\}$ that converges to $R^*$.

In this chapter, it is pointed out that $R^*$ may be very different from $R^0$. This means, that the limit model represents a singularity in the model space.

Now, interpret $P$ as the duration of simulated events, and let $\lim_{k \to 0} P_k = 0$. Thus the model $M(0)$ is a strictly discrete-event model, and it is a singularity in the model space. Consult Sect. 9.3.1 and Fig. 9.3.

# References

1. Barros FJ (1996) The dynamic structure discrete event system specification formalism. Trans Soc Comput Simul 13(1):35–46. The Society for Modeling and Simulation
2. Chow AC (1996) A parallel, hierarchical, modular modeling formalism and its distributed simulator. Trans Soc Comput Simul 13(2):55–67. The Society for Modeling and Simulation
3. Dassisti M, Galantuci LM (2005) Pseudo-fuzzy discrete-event simulation for on-line production control. Comput Ind Eng 49(2):266-286. Elsevier
4. Du X, Ying H, Lin F (2009) Theory of extended fuzzy discrete-event systems. IEEE Trans Fuzzy Syst 17(2):316–328
5. Fishwick PA (1988) Object-oriented Simulation. ACM SIGSIM Simulation Digest, vol 19(2), p 19. New York, NY. https://doi.org/10.1145/47874.1108791
6. Lin F, Ying H (2001) Fuzzy discrete event systems and their observability. In: IFSA world congress and 20th NAFIPS international conference, vol 3, pp 1271–1276, NAFIPS. https://doi.org/10.1109/NAFIPS.2001.943730
7. Novak V, Perfilieva I, Mockor J (1999) Mathematical principles of fuzzy logic. Kluwer Academic, Dodrecht. ISBN/ISSN 0-7923-8595-0
8. O'Keefe RM (1986) The three-phase approach: a comment on "strategy-related characteristics of discrete-event languages and models". Simulation 47(5):208–210. Society for Computer Simulation. https://doi.org/10.1177/003754978604700505
9. Perrone G, Zinno A (2001) Fuzzy discrete event simulation: a new tool for rapid analysis of production systems under vague information. J Intell Manuf 12(3):309–326
10. Raczynski S (2021) Simultaneous events, singularity in the space of models and chicken game. Int J Model Simul Sci Comput 12(4). https://doi.org/10.1142/S179396232140002X
11. Saadawi H, Wainer GA (2010) Rational time-advance DEVS (RTA-DEVS). In: Spring simulation conference (SpringSim10), DEVS symposium. Society for Computer Simulation (SCS)
12. Sadeghi N, Fayek AR, Seresht NG (2016) A fuzzy discrete event simulation framework for construction applications: improving the simulation time advancement. J Constr Eng Manage 152(12), American Society of Civil Engineers. https://doi.org/10.1061/(ASCE)CO.1943-7862.0001195 - See more at: http://ascelibrary.org/doi/abs/10.1061/(ASCE)CO.1943-7862.0001195

13. Santucci JF, Capocchi L (2014) Fuzzy discrete-event systems modeling and simulation. In: SIMUL 2014: the sixth international conference on advances in system simulation. ISBN/ISSN 978-1-61208-371-1
14. Zadeh LA (1965) Fuzzy sets. Inf Control 8(3):338–353
15. Zeigler BP (1987) Hierarchical, modular discrete-event modelling in an object-oriented environment. Simulation 49(5):219–230, SCS
16. Zhang H, Li H, Tam CM (2004) Fuzzy discrete-event simulation for modeling uncertain activity duration. Eng Constr Archit Manag 11(6):426–437

# Chapter 10
# Models and Categories

## 10.1 Introduction: The Language of Categories

In 1942–45, Samuel Eilenberg [2] and Saunders Mac Lane [7] introduced categories, functors, and natural transformations as part of their work in topology, especially algebraic topology (see also [1, 5]). As the concepts of the theory of categories provide a high level of abstracting, the language of the theory may be useful while constructing models and examining their properties.

Recall the definition of the category.

A category is defined by its data and the axioms the data must satisfy. The data are the following:

1. A collection of things called objects. By default, $A$, $B$, $C$,... vary over objects.

2. A collection of things called morphisms, sometimes called arrows.

By default, $f, g, h, ....$ , vary over morphisms.

3. A relation on morphisms and pairs of objects, called typing of the morphisms.

By default, the relation is denoted $f : A \to B$, for morphism $f$ and objects $A$, $B$. In this case, we also say that $A \to B$ is the type of $f$, and that $f$ is a morphism from $A$ to $B$. Object $A$ is called the source of $f$, and object $B$ is its target. In other words, $src f = A$ and $tgt f = B$.

4. A binary partial operation on morphisms, called composition. By default, $f; g$ is the notation of the composition of morphisms $f$ and $g$.

5. For each object $A$ a distinguished morphism, called identity on $A$. By default, $id A$, or id when $A$ is clear from the context denotes the identity on object $A$.

The morphisms of a category must satisfy the following axioms:

A1 $f : A \to B$ and $f : A' \to B' \Rightarrow A = A'$ and $B = B'$ (unique-Type)

A2 $f : A \to B$ and $g : B \to C \Rightarrow f; g : A \to C$ (composition-Type)

A3 $id A : A \to A$ (identity-type)

A4 $(f; g); h = f; (g; h)$ (composition)

A5 $id; f = f = f; id$ (identity)

Note that $f : A \rightarrow A$ does not mean that $f$ is an identity (it is an endomorphism). For example, if $X$ is the space of reals, the function $y(x) = x_3$ maps $X$ to itself, but it is not an identity. Also $f : A \rightarrow B$ and $g : A \rightarrow B$ does not mean that $f = g$.

### 10.1.1   Examples

Following Fokkinga [3, 4]: Each pre-ordered set $(A, \leq)$ can be considered a category, in the following way. The elements $a, b, ....$ of $A$ are the objects of the category and there is a morphism from $a$ to $b$ precisely when $a \leq b$. Formally, the category is defined as follows:

An object is: an element in $A$

a morphism is: a pair $(a, b)$ with $a \leq b$ in $A$

$(a, b) : c \rightarrow d \equiv (a = c) \wedge (b = d)$

$(a, b); (b, c) = (a, c)$

$id \ a = (a, a)$

Another example can be a mechanical system, composed of a set of bodies. The corresponding category $\mathbf{M_0}$ can be defined as follows:

An object of $\mathbf{M_0}$ is the pair $(X(t), t)$ where $X$ is the system state and $t$ stands for the time.

A morphism is a pair $((X(t_1), t_1), (X(t_2), t_2))$ with $t_1 \leq t_2$

$((X(t_1), t_1), (X(t_2), t_2)) : (Y, r) \rightarrow (Z, s) \equiv (Y, r) = (X(t_1), t_1) \wedge (Z, s) = (X(t_2), t_2)$

$id = ((X(t), t), X(t), t))$.

If $\mathbf{M_0}$ is a real physical system, then the morphisms are physical rules of movement (state transitions) from a given time instant to some other, future time instant.

The *functor* is a mapping from one category to another, preserving the categorical structure. It preserves the property of being an object, a morphism, typing, composition, and identity. To be more precise, the definition is as follows (as given in [4]). Let A and B be categories; then a functor from A to B is: a mapping $\mathcal{F}$ that sends objects of $A$ to objects of $B$, and morphisms of $A$ to morphisms of $B$ in such a way that

A6 $\mathcal{F} \ f : \mathcal{F}A \rightarrow_B FB$ whenever $f : A \rightarrow AB$

A7 $\mathcal{F}id_A = id \, F_A$ for each object $A \, in \, A$

A8 $\mathcal{F}(f; g) = \mathcal{F}f; \mathcal{F}g$

The symbol $\rightarrow_A$ denotes a mapping in the category A. The axioms A7 and A8 imply the following: F (f ; . . . ; g) = F f ; . . . ; F g .

Now, consider again the system $\mathbf{M_0}$. What we can do is merely to observe its behavior. This is rather a philosophical assertion telling us that no object or system of the real world can be completely understood and its functioning completely known. So, let us assume that $\mathbf{M_0}$ is a category. If so, we can look for another category $\mathbf{M}$ in such a way that:

A9 The properties of $\mathbf{M}$ are simple enough to be understood

A10 There exists a functor $\mathcal{F} : \mathbf{M_0} \rightarrow \mathbf{M}$.

We will call the category **M** *the model* of $\mathbf{M_0}$, and the functor $\mathcal{F}$ the *operation of modeling*. The model **M** does not have to be unique, so we can look for other, simpler or more sophisticated models for the same physical (or proposed) system. The definition of the validity implies that any valid model must form a category, and the operation of modeling must satisfy the above functor axioms.

For example, define the objects of $\mathbf{M_0}$ as pairs $(X_0(t), t)$, where $X_0$ is the state of a set of particles (3D positions, velocities, body temperature, spacial orientation, spin, etc.) moving in space and subject to gravitational forces in such the way that they do not collide. Then, the model **M** may be a category of material points, the objects of **M** being paired $(X(t), t)$, and the morphisms $f_{a,b}$ of **M** being defined by solutions to a system of differential equations (Newtonian or relativistic) that describe the state transition over a model time interval $[a, b]$. The functor $\mathcal{F}$ maps the real world objects $(X_0(t), t)$ into $(X(t), t)$ taking the positions and velocities of the mass center of each body only. The morphisms of **M** are defined as $f_{1,2} : (X(t_1), t_1) \rightarrow (X(t_2), t_2)$, where $f_{1,2}$ means to solve the model differential equations over $[t_1, t_2]$ with initial condition $X(t_1)$. In the sequel, we will omit the subindices of $f$. If the system $\mathbf{M_0}$ is subject to some external excitation (forces), then this (input) functions should be added to the definition of the morphism.

Note that a consequence of the fact that $\mathcal{F}$ is a functor (see A6) is that the graph of Fig. 10.1 must commute (passing from $X_0(t_1)$ *to* , $X(t_2)$). If it does not, there is something wrong with the definition of the functor. This is a well-known condition of model validity. Bernard Zeigler [10] defines the validity in the same way.

Note that on Fig. 10.1 we have $(X(t), t) = \mathcal{F}(X_0(t), t)$, where the sub 0 means the original (real) system. Remember that the functor $\mathcal{F}$ maps category objects as well as morphisms.

Now, suppose that in the real system the bodies can collide (due to their finite diameter or to the movement itself). In such the case, the graph Fig. 10.1 does not commute, simply because the real transition rule $f_o$ takes into account possible

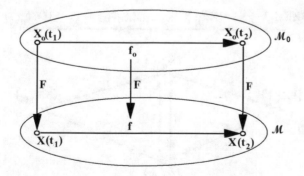

**Fig. 10.1**  Model validity

collisions and the morphism $f$ does not (the model contemplates a movement of
material points only, not the real bodies). In this case, the model is invalid.

For the basic model (the real world), consider the pairs (state, time) as category
objects, and morphisms $f$ as the rules of state-time transition. While constructing the
model, the functor maps *(state, time)* pairs into *(model state, model time)* pairs. This
functor also maps the real system morphism $f$ into the rules of transitions in a discrete
event case, or some differential equation or other rules for the continuous case.
As discussed earlier, the discrete model may provide false results for simultaneous
events. If so, the corresponding graph of Fig. 10.1 does not commute.

Now, take a closer look at simultaneous events. Let category $M$ be a set of pairs
$(X_n, t_n)$, where $X$ is the model state and $t_n$ a time instant. We introduce a weak
order in $M$, with respect to time, namely we define $(X_i, t_i) \le (X_j, t_j)$ iff $t_i \le t_j$. Let
for any two objects $A$ and $B$ in $M$, such $A \le B$ there exists a morphism between
$A$ and $B$. We will interpret morphisms as discrete events. More precisely, an event
morphism between $A$ and $B$ can be expressed as follows:

$f : A \to B \equiv$ *"take the value of X, perform an operation over it and substitute
the result of the value of X in B, take the value of time for $srcf$ add a non-negative
(may be zero) value to it and substitute it as the time of $tgtf$."* We will say that the
event occurs at time equal to the time of $tgtf$.

Consequently, if there are two or more objects with the same value of $t$,
then morphisms must exist between each other. This is the case of simultaneous
events. Figure 10.2 shows a diagram for three events that occur in time instants
$t_n < t_n + e < t_{n+1}$. If e approaches zero, the events $f$ and h have nearly the same
target time. One may suppose, that the model with simultaneous events f and h is the
limit case for $e \to 0$. However, this is not the case. Figure 10.2 shows the diagram
for $e = 0$. As the targets of morphisms $f$ and $g$ have the same value of $t$, there must
exist morphisms between the corresponding nodes, namely $h$ and $i$. The morphism
$h; i$ is not necessarily an identity. As for the morphisms $j$ and $k$ we may assume that

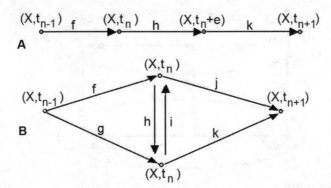

**Fig. 10.2** Two versions of the model. The suffix of $X$ is omitted, supposed to be the same as that
of the time $t$. Identity morphisms are assumed, but not shown

these are identities. Observe the following:

   \* The model changes its structure at when e reaches 0
   \*\* W have $f; h; k = g; i; j$ which means that $f; h = g; i$

## 10.1.2 Simultaneous Events

Consider the following example:

The model $M_d$ has two components $A$ and $B$. Denote the state of the components as $a$ and $b$ (real numbers), respectively.

Denote : model state $x$, a real number, define $x = a + b$.

Result space $= R$ (space of reals), the output $y$ is $y = x$

Let the model includes only two following discrete events:

E1 : add one to the state of component A

E2 : replace the actual value $b$ with $a^2$

The initial condition: $a = 3, b = 0$.

Let the two events execute in time instant $t_1$ and $t_2$ respectively. If $t_1 < t_2$ then the final outcome is equal to 20, and if $t_1 > t_2$ then the outcome is equal to 13.

What happens when $t_1 = t_2$ ? In this case, we have simultaneous events at $t_1$. The two events occur simultaneously in model time. In the real (computer) time, they must be executed in some order. If the order is E1, E2 then the outcome is 20, otherwise it is equal to 13. Which result is valid? Recall that the DEVS (Discrete Event Specification) formalism [6], in the case of multi-component coupled models, provides the formal model component, named *select*. It is a function that embodies the rules employed to choose which of the imminent components (those having the minimum time of next event) is allowed to carry out its next event. However, select is not related to any information provided by the modeled system, at least in this case. Consequently, the modeler must define it. As there are only two options for the execution of the events, the two possible results are equal to 20 for $t_1 < t_2$ or 13 for $t_1 > t_2$. If the modeler has no idea about the selection, then he/she could define the order as random, and take the average value over a set of simulations, which will tend to 16.5. Such outcome, however, never takes place for this model.

The select component may be defined when we have some data about the real system, collected from experiments. However, such approach to the selector definition may be questioned. If we have data about the real system, sufficient to elaborate the select algorithm, we perhaps need no simulation at all.

Relaxing the discrete events to semi-discrete so that each of them has a finite duration (the changes are continuous over some small interval), we can overcome the need for any "select" rule and get rid of the discontinuity and ambiguity caused by the simultaneous events.

In the above example, let both $f$ and $i$ correspond to the event EV1, $g$, and $k$ to the event EV2 (add one to X and take the square of X, respectively). Obviously, $f$; $h$ cannot be equal to $g$; $i$. So, we have an ambiguity at $t_{n+1}$

## 10.2   Conclusion

In fact, we did not prove anything new. This chapter only shows how to express models and their properties in the categorical language, and to encourage researchers to look for new model specifications. The categorical language may be useful when we want to have a higher level of abstraction. Each morphism can be a function, an algorithm, a piece of software, or even a textual specification of state transitions. In the case of the discrete events, we show again (see also [8, 9]) that they form a singularity in the space of models. This may cause problems in model validity.

The theory of categories was created by mathematicians, and most of the books and texts about the theory are full of examples taken from advanced mathematics. Note, however, that the definitions of the basic concepts are not very complicated and need not be always applied to structures of high mathematics. If the knowledge on the category theory were a little bit more propagated in the modeling and simulation community, it might become a good and uniform way to express model specifications.

## 10.3   Questions and Answers

**Question 10.1** What is a *category* (in the theory of categories)?

**Question 10.2** What is a *functor*?

**Question 10.3** How can we interpret the operation od *modeling* in the language of categories?

**Question 10.4** What fact in discrete event modeling we demonstrate here, in terms of the language of categories?

**Answers**

**Answer 10.1** A category is defined by its data and the axioms the data must satisfy. The data are the following:

1. A collection of things called objects. By default, $A$, $B$, $C$, ... vary over objects.
2. A collection of things called *morphisms*, sometimes called arrows.

By default, $f$, $g$, $h$, .... , vary over morphisms.

3. A relation on morphisms and pairs of objects, called typing of the morphisms.

A *morphism* $f$ is a mapping between objects, $f : A \rightarrow B$. For more detail and axioms consult Sect. 10.1.

**Answer 10.2** The *functor* is a mapping from one category to another, preserving the categorical structure. It preserves the property of being an object, a morphism, typing, composition, and identity. Let **A** and **B** be categories. The functor sends objects of **A** to objects of **B**, and morphisms of **A** to morphisms of **B**.

**Answer 10.3** The operation of *modeling* (creating a model) can be treated as a *functor* that operates between the real world and the model.

While constructing the model, the functor maps pairs (*state, time*) into (*model state, model time*) pairs, and the real system interactions into the morphisms of the model (a new category).

**Answer 10.4** Ambiguity problem in simultaneous event modeling, model validity.

# References

1. Asperti A, Longo G (1991) Categories, types and structures. MIT Press. ISBN 0262011255, ftp://ftp.di.ens.fr/pub/users/longo/CategTypesStructures/book.pdf
2. Eilenberg S (1974–76) Automata, languages, and machines (2 vols). Academic Press, New York. ISBN 0-12-234001-9
3. Fokkinga MM (1992) Law and order in algorithmics. PhD thesis, University of Twente, Department of Computer Science, Enschede, The Netherlands
4. Fokkinga MM. A gentle introduction to category theory. http://wwwhome.cs.utwente.nl/~fokkinga/mmf92b.pdf
5. Hoare CAR. Notes on an approach to category theory for computer scientists. In: Broy M (ed) Constructive methods in computing science, pp 245–305. International Summer School directed by F.L. Bauer [et al.], Springer, 1989. NATO Advanced Science Institute Series (Series F: Computer and System Sciences, vol 55)
6. Hu X, Zeigler BP, Hwang MH, Mak E (2007) DEVS systems-theory framework for reusable testing of I/O behaviors in service oriented architectures. In: IEEE international conference on information reuse and integration, Las Vegas, NV, Aug 2007
7. Mac Lane S (1971) Categories for the working mathematician. Graduated texts in mathematics, vol 5. Springer. ISBN 0-387-90036-5
8. Raczynski S (2003) Are discrete models valid? In: VIth conference on computer simulation and industry applications, McLeod Institute for Simulation Sciences, Tijuana, B.C., Mexico, Feb 2003
9. Raczynski S (2000) Alternative mathematical tools for modeling and simulation: metric space of models, uncertainty, differential inclusions and semi-discrete events. In: Proceedings, European simulation symposium ESS2000, Hamburg, Germany, Sept 2000 (Keynote plenary speech)
10. Zeigler BP. Theory of modeling and simulation. Wiley, New York

# Chapter 11
# Fuzzy Time Instants and Time Model

## 11.1 Introduction

In this chapter, we suppose that the time instant is not just a point on the time axis, but it has a finite duration. Inside the interval of such *fuzzy time instant*, the events occur gradually and not in a sharp, discrete moment of time. In the case of time discretization, it is supposed that the duration of a fuzzy time instant is greater than several time steps. The gradual change of the system state is being applied with intensity defined by a probability density function. If the standard deviation of this distribution tends to zero, then the density function approaches Dirac's function, and the changes in the system state occur in discrete (sharp) time instant. The fuzzy time instant may be a better approximation of our perception of time. An example of the behavior of a dynamic second-order system is used to illustrate the model trajectories with fuzzy time instants.

In modeling and simulation of dynamic systems, the concept *time* is always present. In practical applications, the time is understood in an intuitive way, just using the time axis, where one can mark points representing *time instants*. Here, we take a somewhat more detailed look at the time instants, and our perception of what we really understand by "now". Our common perception of the term "now" is somewhat vague, perhaps because the time is running constantly. In this chapter, we permit that the time instant has a finite duration and that the events occur in a somewhat "fuzzy" way. We are not using the concepts of fuzzy set theory, just considering the time instant as fuzzy or "soft" event.

There are many sources where the mechanics of time is considered. Hooft [5] discusses the concept of time-arrow. This concept is closely related to the concept of causality. In the chapter, the time coordinate is subject to ordering, or orientation. This is called the *arrow of time*. The problem of the mechanics of time is frequently focused on the problem of time-reversal. In the conclusion, we read "*this author does not believe that quantum mechanics will be the last and permanent framework for the ultimate laws of nature*". Here, we do not discuss the problem of time-reversal and the implications of quantum mechanics, supposing the time coordinate(s) to have a

© The Author(s), under exclusive license to Springer Nature Switzerland AG 2022
S. Raczynski, *Models for Research and Understanding*, Simulation Foundations,
Methods and Applications, https://doi.org/10.1007/978-3-031-11926-2_11

defined orientation. More discussion on the time properties in quantum mechanics can be found in the book of Muga, Sala and Egusquiza [6].

Craig and Weinstein [3] discuss the multi-dimensional time and the initial-value problems for the wave equation. They show that the Cauchy problem for a higher dimensional case may be ill-posed. This may cause uniqueness problems for the solutions. Let us express here our, perhaps somewhat subjective opinion on such and similar research. Many authors believe that the mechanics of the real world is described by differential equations. This is not exactly true. The differential equations have mostly a nice property of reversibility. Newton's equations of motion can be integrated forward, as well as backward in time. However, in the real world everything is charged with some uncertainty. Introducing such uncertain terms to the movement equation, we obtain a differential inclusion instead of differential equation. The solution of a differential inclusion (the reachable set) is not reversible. Changing the direction of time variable (integrating it backwards in time), we do not obtain the initial point or set, consult Raczynski [8], Raczynski [7]. Note that an uncertain term is not necessarily random. It is rather a *tychastic variable*, see Aubin et al. [1]. For other works on the mechanics of time consult Hilgevoord [4] or Velev [9].

The concept of *fuzzy time instant* consists in considering the time instant as a distribution with small standard deviation that moves in time (over the time axis). It can be any continuous probability distribution. However, in this paper we assume that it is the normal density function, truncated if necessary. The main tenet is that the *events happen gradually during certain time interval*, and not in a sharp time instants.

If the standard deviation of the distribution tends to zero, then the density function approaches the pulse of Dirac, and the time instant becomes as in the conventional approach. Treating the time instant as a distribution, maybe different from zero over the whole time axis, we violate the concept of causality. In our approach, the traditional causality of events does not exist, and the execution of an event may depend on the future. However, this dependence vanishes when the distribution becomes Dirac's pulse, and the causality property recovers its conventional sense. In other words, an event may depend on the future, though the probability of this dependence tends to zero with small standard deviation of the time instant function. If we consider time discretization, as suggested by quantum physics, then the distribution we use is replaced with its discrete approximation. To avoid the violation of the causality, we could define the distribution as equal to zero for all $x \geq t$, $t$ being the actual time. However, in out case we will not do this, allowing the distribution to be non-zero, even over the whole time axis.

## 11.2   The Fuzzy Time Instant

Here, we do not discuss the quantum mechanics concepts of time. We assume that the time may be a continuous or discrete, one-dimensional variable. As stated before, our time instant is a distribution that moves along the time axis. The changes of the state

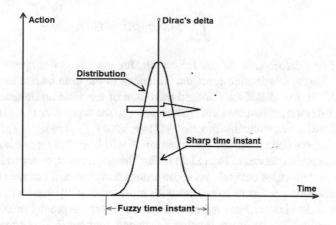

**Fig. 11.1** Sharp and fuzzy time instants

(events) of the system under consideration happen with certain intensity, defined by a density function defined over the interval, see Fig. 11.1.

In other words, the *fuzzy time instant* (FTI) is defined by a distribution function that moves over the time axis. Any probability density function can be used, though this is not a probabilistic approach. Using Dirac's delta function as the distribution, we get a conventional, sharp time instant. In the case of discrete time, we will use a corresponding discrete density function. In this case, we assume that the duration of the FTI is greater than the discretization step. Here, the time instant in the conventional sense will be called *actual time* or *sharp time instant*. It is just a point at the time axis.

Note that the FTIs may overlap. This, and the fact that the events occur over finite time interval raises some causality questions. Here, we assume that the principle of causality is valid in large time intervals (larger than the time instant duration), but inside the FTI it is not. This is perhaps an arbitrary supposition, but anyway, our aim is to learn a "what if" scenario. A modified causality principle can be formulated as follows: The event depends on the system state over the whole interval where the time instant distribution is different from zero. Obviously, if the distribution is Dirac's function, the event depends on the previous time instants (we discard exactly simultaneous events). If the standard deviation of the distribution approaches zero, then the probability that the event depends on the future states also tends to zero.

First, let us see what is our perception or observation about the state of the world, or a part of it referred to as a system. The (real) system state $x$ is supposed to be a function of time. It can be a scalar- or vector-valued function. Then, our observation is given by the following expression.

Consider any function of the time and state, $f(x, t)$. The observed value of $f$, denoted as $\mathcal{F}(x, t)$, is given by the following equation.

$$\mathcal{F}(x, t) = \int_{-\infty}^{\infty} f(x, \tau) p(\tau - t) d\tau, \tag{11.1}$$

where $p(*)$ is a distribution. In the following, for the sake of simplicity, we use the normal density distribution function, truncated to the time instant interval, and normalized. Observe that if the standard deviation of the distribution tends to zero, then the distribution approaches Dirac's function, and the expression (11.1) reduces to $x(t)$. Recall that for an integrable function $y(t)$ we have $\int_{-\infty}^{\infty} y(\tau)\delta(\tau - t)d\tau = y(t)$, where $\delta$ is Dirac's delta function. The function $\mathcal{F}$ will be called *fuzzy observation*.

One of the consequences of Eq. (11.1) is that the observed state depends not only on the past system states but includes some values from the near future. This means that our perception permits us to predict with more probability what will happen in the very small time instant. For example, a football player can predict where his body will be located within the future fraction of second, and hardly can know where it will be after five minutes.

In the case of discrete time, the system state is a sequence $x_n = \{x_0, x_1, x_2, ....\}$, and the discrete version of the expression (11.1) is as follows.

$$\mathcal{F}_k(x) = \sum_{m=k-n}^{k+n} f(x_m, hm) p_{m-k}, \tag{11.2}$$

where $k$ is the number of the discrete time step, and $h$ is the time step. Here, $n$ defines the interval where $p$ is defined, that is, $p_k > 0$ for all $-n \leq k \leq n$, zero otherwise.

Looking at Eq. (11.1) we can see that inside the (FTI), the observation $\mathcal{F}$ depends on all states along the FTI (remember that $f$ depends on $x(t)$). If the movement is described by differential equations, then this set of states is changed according to the derivatives that, in consequence, depend on the value of the observation over the whole FTI. This makes it difficult to integrate the model trajectory because the rate of change at the moment depends also on some future values of the model state. Consequently, it is necessary to apply an iterative algorithm, as explained in the next section.

The *fuzzy derivative* $\mathcal{D}$ of $x(t)$, is defined as follows.

$$\mathcal{D}(t) = \lim_{h \to 0} \frac{\int_{-\infty}^{\infty} x(\tau) \left(p(\tau - t) - p(\tau - t + h)\right) d\tau}{h} \tag{11.3}$$

In the discrete time version, this converts in the following.

$$\mathcal{D}_k = \frac{1}{h} \sum_{m=k-n-1}^{k+n} x_m (p_{m-k} - p_{m-k+1}), \tag{11.4}$$

where $k$ is the time step number, $p_k$ is the discrete density function defined over $[n-1, n+1]$, $p_{n-1} = p_{n+1} = 0$, $h$ is the discrete time step.

Consider the following equation of movement (supposing that the movement is described by an ordinary differential equation):

$$\frac{dx}{dt} = f(x, t) \tag{11.5}$$

In the conventional Euler integration procedure, the change of the model state is done in consecutive time steps: $x_k = x_{k-1} + h f_{k-1}$, where $h$ is the time step interval and $f$ is the derivative. In the FTI version, this event executes during the fuzzy time interval, and the change of state is done according to the fuzzy observation $\mathcal{F}$. Namely, it is applied to the state vector over the fuzzy time instant duration, with the weight defined by the distribution $p$ (see Eq. (11.1))

The problem is that in the FTI, the change of $x_k$ depends not only on the past, but also on some future values of $x$ (see the definition of $\mathcal{F}$, (11.1) and (11.2)). This is the consequence of the lack of causality principle inside the FTI. In this case, we cannot integrate the model trajectory using the simple Euler's method, starting from an initial condition and advancing in time. We cannot change the sum limits so that no future values of $x$ will be needed because would introduce an additional, artificial time delay. We must look for a global solution for all the $2MN$ values of $x$, where $M$ is the total number of time steps of the trajectory, and $N$ is the model dimensionality.

Below, $x$ and $x^*$ are vectors of $N$ components, and $[-n, n]$ is the interval where the density function is defined. A heuristic algorithm may be as follows.

**Algorithm A**

1. Define initial trajectory $x_k = 0$ for $k = -n, ..., M$ (the whole trajectory of $M + n$ discrete values). The values of $x_k$ for $k < 0$ can be set equal to zero.

2. Store the actual trajectory in the array $x_1^*, x_2^*, ..., x_M^*$.

3. Calculate $\mathcal{F}_k, k = 1, .., M$ according to (11.2)

4. Integrate system trajectory using $\mathcal{F}$, store the trajectory in $x_k, k = 1, ..., M$. The integration procedure applies the increment $\mathcal{F}_{k-1}$ to all $x_j, j = k - n/2 .... k + n/2$, each increment with weight $p_{j-k}$.

5. Calculate $z_k = (1 - q)x_k + qx_k^*, \ k = 1, ..., M, \ 0 < q \le 1$, ($x_k^*$ is the state taken from the previous iteration).

6. Substitute $x_k$ with $z_k$ for $k = 1, .., M$

7. Repeat steps 2 to 6 until the solution $x_k$ does not change.

Again, it is easy to see that if $p$ tends to Dirac's function, then the expression in (11.3) reduces to the conventional derivative of $x$.

Note that in this algorithm we apply a "double averaging," calculating the fuzzy observation $\mathcal{F}$, and then applying the necessary changes of model state inside the fuzzy time instant and not in the current point on the time axis. This is not a unique algorithm for the implementation of fuzzy time instant dynamics. We can use an exact observation, as well as a fuzzy observation applied in the strict time instant. What is required is the fuzzy observation or fuzzy state changes to be applied in

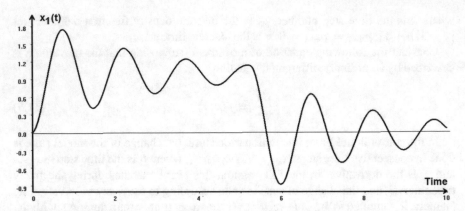

**Fig. 11.2**  Simple simulation of system (11.6)

such a way that the results converge to the conventional state integration when the standard deviation of the distribution $p$ tends to zero.

## 11.2.1  Example

Consider a second-order dynamic system given by the following state equations.

$$\begin{cases} \dfrac{dx_1}{dt} = x_2(t) \\ \dfrac{dx_2(t)}{dt} = A(u(t) - x_1(t)) - Bx_2(t), \end{cases} \tag{11.6}$$

where $A$ and $B$ are constants, and $u(t)$ is an external control function. Let $A = 20$, $B = 0.8$ and $u(t) = 1$ for $t \leq 5$, $u(t) = 0$ for $t > 5$. Figure 11.2 shows a simple simulation of the system movement over time interval $J = [0, 10]$ (variable $x_1$ shown). Here, we assume discrete time, 10000 time steps over the interval $J$. So, the integration is carried out by simple Euler method, that provides the sequence $x_n = x_0, x_1, x_2, .... x_M$, M = 10000.

Now let us see how the use of fuzzy observation and gradual event execution influence the model trajectory.

Consider the trajectory integration with fuzzy time instants, non-zero standard deviation, and discrete time. Now, the model state $x_k$ changes not in the one, discrete time instant, but the change occurs gradually when the time instant passes over the time equal to $k$. The rate of change $\mathcal{F}_k$ at time $k$ is given by Eq. (11.2).

We use the discrete density $p_n$ defined over the interval $-250 \leq n \leq 250$ that corresponds to the interval $[-0.25, 0.25]$ of continuous time with time step equal to 0.001. The density function is derived from the normal distribution with standard

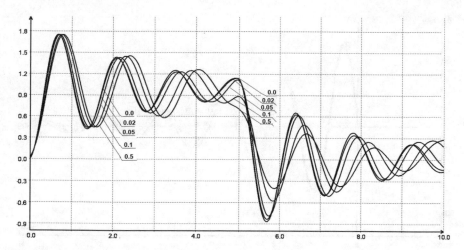

**Fig. 11.3** Trajectory $x_1(t)$ for different values of standard deviation of function $p$. The plot with value 0 is the result of simple simulation with sharp time instants

deviation $s$, centered in $n = 0$ truncated and normalized. So, the distribution value at the point $time = hi$ is equal to $p_{i-k}$, where $k$ is the discrete time where the fuzzy time instant is actually located.

In the above example, the solution provided by the first iteration is unstable, and $x$ grows or decreases rapidly for greater values of $k$. Generating the plot of $x_k$ it can be seen how the stable solution propagates toward greater $k$. Finally, we get a stable solution that no longer changes in the consecutive steps of the algorithm. It is not the topic of this paper to prove that the algorithm converges for other models. We just look for the solution to our problem to illustrate the behavior of the model with FTIs. The algorithm does not converge for $q = 0$. A stable solution was obtained for $q = 0.5$.

Figure 11.3 shows the plot of $x_{1,k}$ (the first component of $x$, time step $k$) obtained with different values of the standard deviation of $p$. The curve with deviation zero is the result of the simple simulation with sharp time instants, like in Fig. 11.2. These results are similar to the trajectories that can be obtained by a smoothing algorithm applied to the simple simulation. However, what we are doing is not smoothing a curve, but simulating the model behavior with fuzzy time instants. The standard deviation values correspond to the original normal distribution before truncating and normalizing. The values are expressed in the continuous time units (the whole simulated interval is equal to [0,10]).

In Fig. 11.4 we can see the model response to Dirac's pulse applied at $time = 2$, for fuzzy time instant with standard deviation 0.05 and 0 (simple simulation). Observe that the model with fuzzy time instant is able to "predict" the pulse, and the response starts slightly before the excitation. Again, this is due to the lack of causality within the fuzzy time instant.

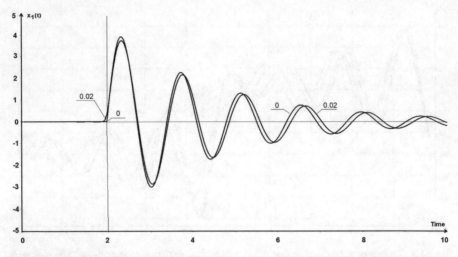

**Fig. 11.4** Model response to Dirac's pulse at time equal to 2, for standard deviation 0.02 and 0

## 11.3  Conclusion

We consider a generalization of the concept of the time instant. Instead of the sharp time instant defined as a point at the time axis, we define the fuzzy time instant as a (maybe narrow) distribution density function that scans the time axis. Such a definition of the time instant perhaps fits better to our intuitive perception of the time and the concept of "now". Also note that if an event is related to energy flow, then the model with fuzzy time instants may describe better the system dynamics.

We assume that the observations of the system state variables are the weighted averages due to a distribution function. Moreover, it is assumed that the events, like the changes in the model state, occur gradually inside the fuzzy time instant. It can be seen that the fuzzy time instants and the observation results converge to the conventional sharp time instants, if the distribution function tends to the delta function of Dirac. The model events occur within the fuzzy time instant with finite duration and not in a sharp moment of time. One of the consequences of the fuzzy time instant is the violation (or redefinition) of the principle of causality. The conventional causality does not work inside the FTI. However, if we redefine the causality in a probabilistic way, it tends to the strict causality definition when the FTI distribution function tends to Dirac's pulse. An example of oscillator dynamics presented here depicts the difficulties and the way the system trajectories can be calculated.

# References

1. Aubin JP (2013) Tychastic viability: a mathematical approach to time and uncertainty. 61(3):329–340. https://doi.org/10.1007/s10441-013-9194-4
2. Aubin JP, Cellina A (1984) Differential inclusions. Springer, Berlin
3. Craig W, Weinstein S (2009) On determinism and well-posedness in multiple time dimensions. 465:3023–3046. https://doi.org/10.1098/rspa.2009.0097
4. Hilgevoord J (2002) Time in quantum mechanics. Am J Phys 70(301). https://doi.org/10.1119/1.1430697
5. Hooft G (2018) Time, the arrow of time, and quantum mechanics. Front Phys. https://doi.org/10.3389/fphy.2018.00081, ISBN/ISSN 2296-424X
6. Muga G, Sala Mayato R, Egusquiza I (2008) Time in quantum mechanics. Springer, Berlin Heidelberg. https://doi.org/10.1007/978-3-540-73473-4, ISBN: 978-3-540-73473-4
7. Raczynski S (2019) Differential inclusions in modeling and simulation. Brown Walker Press. ISBN: 1627347429
8. Raczynski S (2011) Uncertainty, dualism and inverse reachable sets. Int J Simul Model 10(1):38–45. ISBN/ISSN: 1726-4529
9. Velev MV (2012) Relativistic mechanics in multiple time dimensions. Phys Essays, Phys Essays Publ 25(3):403–438. https://doi.org/10.4006/0836-1398-25.3.403

# Chapter 12
# Uncertain Future, Reversibility and the Fifth Dimension

## 12.1 Introduction

The contents of this chapter are not science fiction, though some fictitious elements can be found here. We treat some, rather abstract, models of a real world that can hardly be validated but may serve as a cognitive insight and new paradigms in modeling. In the following sections, we discuss a model with feedback from the future (the *ideal predictor*), the problem of reversibility of model trajectories (as well as the events of the real world), and a model of the universe, encapsulated in a ball with finite radius. In fact, the *encapsulated universe* is an example of a useless model which, however, may have some cognitive elements, and may stimulate our imagination. According to this model, all the universe we can observe is enclosed in an open unit sphere. The *local observer* (like us) cannot see the sphere limits and the non-linearity of the metrics inside the sphere. Once the sphere has finite radius, the problem arises if a moving particle can go out of the sphere, and what happens if it does. This would be a *trip behind the infinity*.

## 12.2 Uncertain Future

This section treats on the trajectory calculations for dynamic model with positive time-shift (ideal predictor). We replace the element with unknown future model state with a set that represents the uncertainty of the future. This leads to a differential inclusion. The differential inclusion solver program provides the solution to this inclusion, in the form of a reachable set in the time-state space. Then, an iterative process is proposed that converges to a single trajectory, being the solution to the original problem with the positive time-shift. Some examples of linear and non-linear models with ideal predictors are provided. This treatment of the positive time-shift is different from the known predictor algorithms. We use the differential inclusion solver to treat the problem, instead of differential equations.

© The Author(s), under exclusive license to Springer Nature Switzerland AG 2022    247
S. Raczynski, *Models for Research and Understanding*, Simulation Foundations, Methods and Applications, https://doi.org/10.1007/978-3-031-11926-2_12

If we suppose that our world is causal, then we should discard the possibility of traveling to the future, going back, and making decisions or actions. The decisions are taken according to the information taken from the future. There are a lot of examples of the paradoxes that arise from time traveling. Let recall only one, perhaps the most illustrative. Imagine, that young Beethoven has traveled to the future and went to a concert where his Fifth Symphony is being played. Then, he returns to the original time. As he has the perfect musical memory, he puts the music he heard into the musical score and publishes it. Thus, the *Fifth Symphony has never been composed*.

This chapter is not science fiction. We discuss just the possibility of simulating an ideal predictor (that, in fact, contradicts the principle of causality). Note that we do not deal with any known prediction algorithms used in decision-making processes, in marketing, economy, stock market, control, and other fields. Such predictors provide an approximate, perhaps more probable scenario for future system behavior, but they are not ideal predictors. By the *ideal predictor* we mean an object that gives us the exact information about the future system state.

The literature, as well the WEB, is full of articles and other publications on time-traveling. Most of them belong to the field of science fiction, so we will not discuss them here. From some more serious publications let us mention Nahin [13], where we can find remarks on time concepts in quantum physics and more references.

In the field of system dynamics and control theory, there are many publications that deal with time-delayed systems. These considerations are more realistic because we can, to some extent, have the information about the history of the system under consideration. This information may be exact or charged with some uncertainty. On the other hand, the future system states are more uncertain, and the uncertainty grows with the time distance to the supposed future system state.

We restrict our problem to the continuous ODE models. Consider the equation.

$$x' = f(t, x(t), x(t + r)), \tag{12.1}$$

where the prime mark stands for time-differentiation. Here, $x$ is a point in the one- or multi-dimensional state space, $r$ is the time-shift (negative for time-delay systems), and $f$ is a vector-valued function. Another, linear version of time-shift system is frequently given by the following equation

$$x' = \mathbf{A}x(t) + \mathbf{B}x(t + r), \tag{12.2}$$

where $\mathbf{A}$ and $\mathbf{B}$ are matrices.

If $r$ is less or equal to zero, there is no problem with simulating the above models, with given initial conditions for $x$. If no analytical solution can be found, we can apply any of the known numerical methods for ordinary differential equations. The presence of the delayed argument requires the past model trajectory to be stored. Having such a history file, we can retrieve the data necessary to advance with the solution in time. If the time-shift $r$ is positive, integration of model trajectory is difficult. Note that we cannot simply reverse the time axis and go backwards, converting the positive time-shift into a delay because we don't have the final state conditions.

The second argument of the function $f$ of (12.1) is unknown and uncertain. Now, let us treat this argument not as a point in the state space, but a set $V$, as below.

$$x' = f(t, x(t), V(t)), \tag{12.3}$$

where $V$ is a set, maybe changing in time. This set represents the uncertain future. This way, our model (12.1) takes the form of a differential inclusion, instead of a differential equation (consult [3]):

$$x' \in F(t, x(t)). \tag{12.4}$$

Here, $F$ is a set-valued function, defined as follows:

$$F(t, x) = \{z : z = f(x, u, t), \ u \in V(t)\} \tag{12.5}$$

Consequently, our problem is to solve a differential inclusion, instead of differential equation. In practice, we can assess the possible limits of $V$. The solution to a differential inclusion is a set, called *reachable or attainable set*, discussed in Chap. 3. Observe that the equations of the original problem (12.1) include undefined elements. In turn, in the corresponding differential inclusion (12.4) nothing is undefined. Instead of uncertain parameter, we have a well-defined set. Differential inclusions may be used as an alternative to stochastic and probabilistic approach to uncertainty. Instead of stochastic, we can use *tychastic* variables, which values are uncertain, but not random, see Aubin [4].

Figure 12.1 shows a rough explanation of our problem. It is supposed that we have exact information about the present model state $x$ (it also could be questionable). The future states are uncertain (gray region). In the figure, there is also assumed some uncertainty of the past states.

While using differential inclusions, the problem with time delay is not very different from the problem of the ideal predictor. In both cases, we need to solve the corresponding differential inclusion. In this section, we consider a somewhat abstract problem of systems with positive time-shift, and we don't care about the causality principle. We are just looking for a method to solve differential equations with a positive time shift. Some examples are given in further on.

## 12.3 Differential Inclusion Solver

The overview of differential inclusions has been done in Chap. 3. Here, let us recall the applications of the differential inclusion solver (DI solver) (conult also Sect. 3.6).

While working with differential equations, one can find a huge number of numerical methods and software. On the other hand, for the DIs there is nearly nothing that

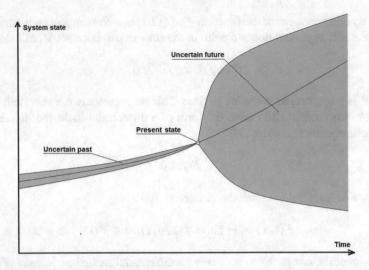

**Fig. 12.1** System trajectory, uncertain past, and future

could help a simulationist. The DI Solver is the result of an attempt to fill this gap.
The basic version of the DI solver is not new. It was published in 2002, see [17].
Here, we only recall the algorithm. In Chap. 3, a new version of the solver and some
new applications to the functional sensitivity are described.

The solver has been coded in the Embarcadero Delphi. A limited stand-alone
".exe" version of the solver is available. It should be emphasized that our main goal
is the RS determination and not optimization. The DI solver and the present problem
statement should not be confused with the differential inclusions method used in the
optimal control problems.

To avoid repetitions from earlier articles, we will not discuss here the algorithm
of the DI Solver. The basic version of the algorithm has been published in [17]. In
few words, the Di Solver generates a series of DI trajectories that *scan the boundary,
and not the interior* of the reachable set. The differential inclusion is derived from
the model state Eq. (12.6) that includes the vector of uncertain parameters $u =
(u_1, u_2, ...u_m.)$.

$$\frac{dx}{dt} = f(x(t), u(t), t) \ t \in I = [0, T], \ T > 0 \tag{12.6}$$

Vector $u$ belongs to a given set of restrictions $C$. When $u$ scans the interior of the
set $C$, the right-hand side of (12.6) scans the set $F$, being the right-hand-side of the
corresponding differential inclusion (12.4).

The solver algorithm uses some results from the optimal control theory (Pontryagin [15]). From Pontryagin's Principle of Maximum it is known that each model
trajectory that reaches a point on the boundary of the reachable set at the final simulation time, must belong to the boundary of this set for all earlier time instants. Consult

Pontryagin [15] and Lee and Markus [10]. Moreover, a boundary-scanning trajectory must satisfy the Jacobi-Hamilton equations (the necessary condition). These equations involve a vector of auxiliary variables $p = (p_1, p_2, p_3, ..., p_n)$. In the optimal control, the problem is that we have the initial conditions for the state vector $x$, but the conditions for the vector $p$ are given at the end of the trajectory, for $t = T$. If no analytic solution is available, then we must use an iterative procedure to solve the so-called *two-point-boundary-value* (TPBV) problem. In our case, we are in better situation. Observe that starting with the given initial condition for $x$ and with any initial condition for $p$, we obtain a trajectory that scans the boundary of the reachable set. This means that we do not have to solve the TPBV problem. The algorithm generates a series of trajectories with randomly generated initial conditions $p(0)$. After integrating a sufficient number of trajectories, we can see the shape of the reachable set boundary. In other words, we are looking for the mapping $MP : p(0) \rightarrow RS(T)$, where $RS(T)$ is the boundary of the reachable set at $t = T$. The problem is that the mapping MP may be extremely irregular even for a simple linear model. The solver algorithm uses certain heuristic procedure to avoid the "holes" of the final image.

Somebody might suppose that by generating a number of trajectories randomly from inside of $F$, we can cover the inside of the reachable set with sufficient density, and then estimate its shape. Unfortunately, this is not the case, even if we select only points from the boundary of the set $F$. Simulation experiments show that even in very simple cases the set of trajectories provided by such *primitive shooting* (using any density function) is concentrated in some small region inside the reachable set and does not approach its boundary. Anyway, it is an error to explore the interior of the reachable set. The DI solver used here explores the boundary of the set, and not the interior. With a sufficient number of generated trajectories, the shape of the reachable set can be estimated with reasonable accuracy.

## 12.4   Solving the Ideal Predictor Problem. Feedback From the Future

Suppose that the present state of the system is known (see Fig. 12.1), and the future states are uncertain, but belong to a given (maybe big) permissible set. Our model is given in the form of differential inclusion (12.4). Using the DI solver, we can solve the inclusion and obtain the reachable set (RS) for a future time interval. The resulting RS can be greater or smaller than the estimate of the limits of an uncertain future state. As the initial estimate is normally given as a big set, it is probable that the RS we obtain is smaller than the initial uncertainty estimate. In many practical applications, this is the case because many physical systems have some limiting elements like saturation and delimiters. Now, we replace the initial uncertainty estimate with the RS provided by the last run of the solver and solve the inclusion again. Repeating these steps, we get an iterative process that can converge or not to a single function

$x(t)$. If the process converges, then we obtain a very narrow RS, which is an estimate
of the solution to the original problem with an ideal predictor.

### 12.4.1   Example 1: A Linear Model

Consider the following model:

$$\begin{cases} x_1' = x_2 - 0.07x_1(t + 0.6) \\ x_2' = 10 - x_1(t) - 0.35x_2(t + 0.6) \end{cases} \qquad (12.7)$$

with initial condition $x_1(0) = x_2(0) = 0$, final simulation time equal to 6.

Figure 12.2 shows the 3D image of the boundary surface of the reachable set for
model (12.7). In this run of the solver, the initial uncertainty of the state variables was
defined as a permissible rectangle that delimits both variables to the interval $\pm 100$.
The range of the final state resulted to be even greater, with $-165 < x_1 < 165$ and
$-161 < x_2 < 156$, approximately. During the calculation process, the actual ranges
for the two variables have been stored. In the next iteration, the stored range values
are used to calculate the next approximation of the RS. In this case, this iterative

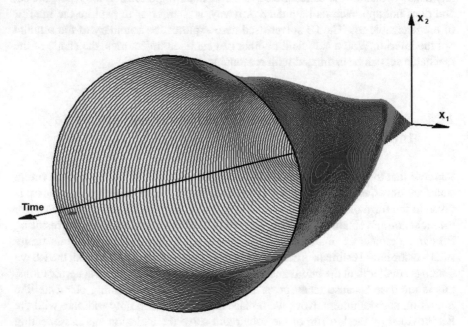

**Fig. 12.2** The 3D image of the reachable set for model (8)

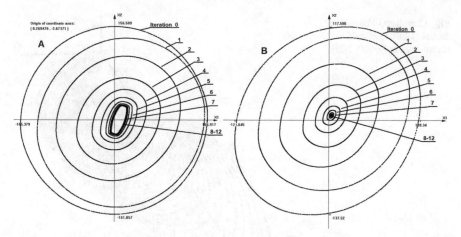

**Fig. 12.3**  Comparison of the RS contours for final time instant (A) and time $= 5$ (B), consecutive iterations

process produces a sequence of reachable sets that converge the solution to the original time-shift problem (12.1).

In Fig. 12.3 we can see the comparison of the consecutive reachable sets for the final time equal to 6 (part A) and 5 (part B). Note that for the final interval $[t - r, t]$ we have no range parameters stored for $x(t + r)$, so the set $V$ is taken equal to that of the first iteration ($\pm 100$ rectangle). This is why the contour of the final reachable set is always finite and does not converge to one point. The first contour (iteration 0) looks somewhat different from the next contours because in the first iteration the set $V$ is equal for all time steps, and in the consecutive iterations it is different for each time-step (stored before). From the part B of Fig. 12.3, we can see that the convergence of the iteration process is quite good.

Figure 12.4 shows the comparison of reachable sets for iterations number 5 and 10. The images have the same scale as that of Fig. 12.3. The convergence of the process is clearly seen.

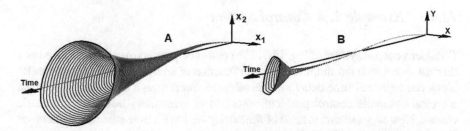

**Fig. 12.4**  Comparison of reachable sets from iteration number 5 (A) and iteration 10 (B)

**Fig. 12.5** Reachable set for
model (12.8), uncertain
future state in [−100, 100]

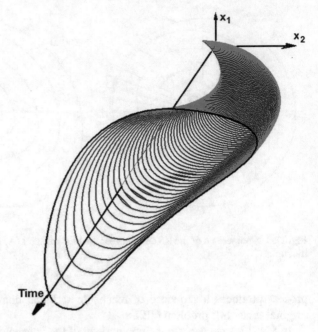

### 12.4.2   Example 2: A Non-Linear Model

The system equations are as follows:

$$\begin{cases} x_1' = x_2 + 0.1x_1(t+r) \\ x_2' = 20 - x_1(t) - 0.2x_2(t) - 0.024\,x_2^2(t) \end{cases} \tag{12.8}$$

Here, $r = 0.35$, final simulation time equal to 5. Figure 12.6 shows the convergence of the iterative process, like in example 1.

### 12.4.3   Example 3: A Control System

Consider a control system of Fig. 12.7. The controlled process may be interpreted as a thermal object with the simplified transfer function of second order. The "time-shift" block can represent time delay or time advance. Supposing a time delay we obtain a typical academic control problem described in elementary books on automatic control. Now suppose that instead of time delay we have a time advance element, an "ideal predictor," the "feedback from the future."

Here, $\Theta$ is the process output that may be, for example, the temperature to be controlled. $u$ is the set point (desired temperature), $K$ is the controller gain, and $y$

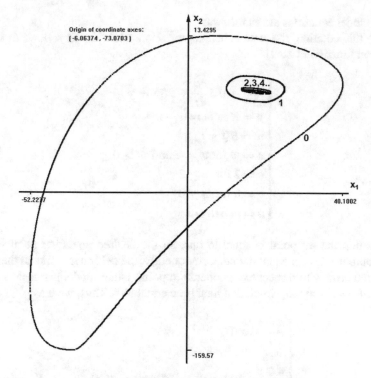

**Fig. 12.6** The contours of the reachable set for model (12.8), consecutive iterations

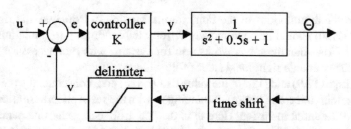

**Fig. 12.7** A control system with ideal predictor

is the controller output signal. The actuator delimiter is described by the function $v = g(w)$. Let denote: $x_1 = \Theta$ and $x_2 = d\Theta/dt$. The output signal of the time-shift element is equal to $x_1(t + r)$, where $t$ is the time and $r$ is a positive constant.

The model equations are as follows:

With this notation, the state equations for this system are as follows (a simple saturation function is used).

$$
\begin{cases}
\dfrac{d^2\theta}{dt^2} + 0.5\dfrac{d\theta}{dt} + \theta(t) = y \\[2mm]
y = Ke, \quad e = u - v \\[1mm]
w = \theta(t + r) \\[1mm]
v = w \text{ for } w < 2 \text{ and } w > 0 \\[1mm]
v = 2 \text{ for } w \geq 2 \\[1mm]
v = 0 \text{ for } w \leq 0 \\[1mm]
u = const = 1
\end{cases}
\tag{12.9}
$$

Note that the set point is equal to one, so the desired operation point for the model output $x$ is also equal to one. Consequently, the delimiter works in the range 0–2, symmetric with respect to this operation point. Other model parameters are as follows. $K = 2$, $r = 0.5$, simulation final time equal to 8. Thus, we have

$$
\begin{cases}
\dfrac{dx_1}{dt} = x_2(t) \\[4mm]
\dfrac{dx_2}{dt} = K(1 - v) - 0.5x_2(t) - x_1(t)
\end{cases}
\tag{12.10}
$$

Due to the assumptions of the Pontryagins Maximum Principle, the right-hand sides of (12.10) should be continuously differentiable. From the simulation experiments, it follows that the solver works also for systems with (not necessarily continuously) differentiable right-hand sides of the state equations.

In the Eqs. (12.9) and (12.10) the value of $x(t + r)$ and, consequently, the value of $v$ are uncertain (uncertain future). Replacing $w$ with the set of all its permitted values, we get a differential inclusion. Here $w$ is the "control" parameter that parametrizes the set of the DI right-hand side. After introducing the above equations to the DI solver, we obtain the following results. Figure 12.8 shows the system reachable set for the initial uncertainty range. The resulting limits of $w$ are stored for all time steps. In the next iteration, these limits are used, and so on.

In this case, the convergence of the iterative process is also quite good. After several iteration we obtain an approximation of the trajectory that satisfies Eq. (12.10). Figures 12.9 and 12.10 show the shape of the reachable set for iterations 14 and 40, respectively.

In this experiment and in other similar situations the algorithm converges quite well. However, there are situations when it does not. The convergence is out of the scope of this chapter and could be a good subject for more theoretical considerations. Note that this is not the problem of the convergence of the DI solver. Here, the solver is used as a part (one step) of a bigger algorithm. If the convergence occurs, we obtain

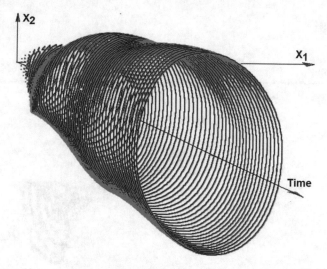

**Fig. 12.8** Reachable set for the initial future uncertainty

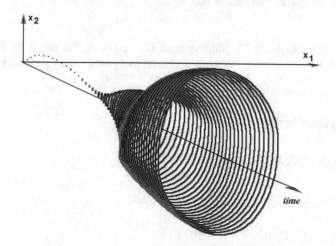

**Fig. 12.9** Reachable set after iteration 14

one and maybe unique trajectory that satisfies the general rules of movement on a system with an ideal predictor. This means that if the procedure converges to one trajectory, the system with "traveling-to-the-future-element" may be stable. But, this is a rather abstract and philosophical question.

It can be shown that a small external disturbance added to the feedback signal affects considerably the convergence. It also can be seen that the stability of the model itself is necessary for the convergence. Running the same model with, for example, $K = 5$, the algorithm does not converge to any single trajectory. As for the disturbances in control systems, if we treat them as uncertainty instead of stochastic

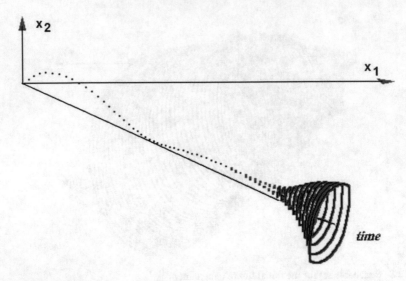

**Fig. 12.10** Reachable set after iteration 40

functions, we again get a DI. Solving such DI we obtain the range for the output variables. This may be useful in such problems as system safety and robustness

## 12.5  Reversibility

Consider a dynamic ODE model, as follows:

$$\frac{dx}{dt} = f(x(t), t) \tag{12.11}$$

with initial condition $x(0) = 0, t \in [0, T]$.

The problem for an ODE model is: Given the final state of a dynamic system, what was the initial state that originated it? For the ODE model (12.11) we can simply go backwards in time, from $T$ to 0. This means that we solve the equation:

$$\frac{dx}{ds} = -f(x(s), s), \tag{12.12}$$

where $x$ may be a scalar or a vector-valued model state.

Theoretically, this works, and we will get the original, initial model state. Unfortunately, this works only for analytical solutions. The reversibility of the numerical solution, as well as a hypothetical reversibility of the real world, do not take place, for a simple reason: **the uncertainty**.

It might appear, that starting from a given final state, and using a sophisticated numerical method, we can obtain the exact initial state. However, this is not true. The reason is very simple: in a digital computer, even a simplest operations on real numbers (float or double type) like *addition and subtracting, are not reversible* Try the following. Let $x = 1.0$ and $y = 1.0/3.0$. Perform the addition $x = x + y$, 200 times. Then, execute $x = x - y$ also 200 times, and print the initial and final value of $x$ with 10-digit precision. We get (C++Builder used $x$, $y$ of type float)

Initial $x = 1.0000000000$

final $x = 1.0000032187$

If these, elemental operations are irreversible, then any complicated numerical method that calculates thousands of integration steps with huge number of arithmetic operation cannot be reversible. The size of the error does not matter. The uncertainty in dynamic models, as well as in the real world always exists. Inside a digital computer, this may be the imperfection of the arithmetics, and in the real world the natural uncertainty of things.

### 12.5.1 Irreversibility of Differential Inclusions

Consider a differential inclusion in the most general form, with necessary regularity assumptions (see Chap. 3).

$$\frac{dx}{dt} \in F(x, t), \tag{12.13}$$

with initial condition $x \in X_0$, where $X_0$ is the initial set. Let (12.13) be considered over the interval $[t_0, t_1]$, $t_0 z t_1$.

Now consider a similar inverse problem for the model (12.13) (a dynamic system with uncertainty): Given the final reachable set, what was the initial set that originated it? In this case, we cannot just go backward in time. Indeed, consider, for example, the inclusion:

$$\frac{dx}{dt} \in [-1, 1], \tag{12.14}$$

Starting from the initial point $x = 0$, $t = 0$ and calculating the $RS$ at $t = 1$ we get $RS(0, 1) = [-1, 1]$. Now, if we start from $t = 1$ and from the final set $[-1, 1]$, and go backwards (inverting the sign of the derivative), we get the set (interval) $[-2, 2]$ instead of the original set (point) $(0, 0)$. It is clear that the $RS$ must grow, because the uncertainty remains the same for the inverse problem. Figure 12.11 illustrates this in two-dimensional case. The reachable set obtained by inverting the sign of the derivatives (going backward in time) is denoted as $BRS$, and $B = RS(A, t_1)$.

In the general case of the differential inclusion (12.13), over a fixed time interval, the solution to the inverse problem should give us the initial set $A$ (see Fig. 12.11). Denote this solution by $IRS$. So, it should be $IRS(RS(A, t_1), t_0) = A$. In other

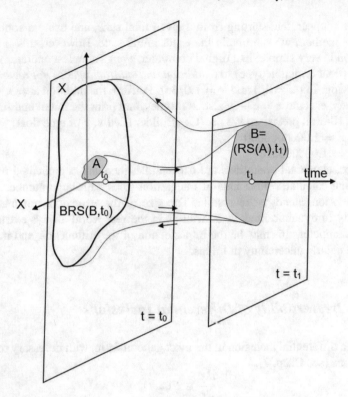

**Fig. 12.11**   Initial set A, reachable set RS and the backward reachable set BRS

words, we must express the set $A$ in terms of the reachable set at $t = t_1$. The second parameter of $IRS$ (as well as of $BRS$) is the target time instant.

In [16] the following corollary has been prooved:

**Corollary 1.** *The $IRS(B, t_0)$ is the intersection of all the $BRSs$ going out of the points of B as initial sets at $t_1$, backwards in time.*

The above corollary does not mean that the differential inclusions are reversible. This is a theoretic result. A practical application of the corollary may be difficult because it requires an infinite number of backward reachable sets to be calculated. This leads to numerically untractable problem.

In [16] there is a remark on a possible generalization. Namely, if we define the reachable set as a set of all possible future model states without referring to differential inclusions and the strictly mathematical definition of the IDs and model trajectory, the corollary 1 remains valid. This means that it may work also for discrete time-space models and any other model dynamics where the reachable set is well defined. The irreversibility of the trajectories of real systems is illustrated in Fig. 12.12. The reachable set is not necessarily a solution to any differential equation or inclusion.

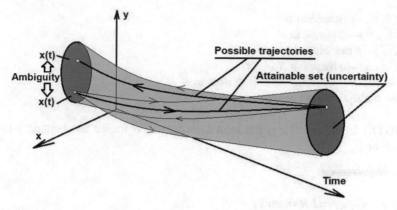

**Fig. 12.12** Irreversible trajectories in presence of uncertainty

## 12.6 Encapsulated Universe and the Fifth Dimension

Let us discuss a fictitious world, encapsulated in a ball. This can be treated as a *fiction*, but it is not what we commonly understand as *science fiction*. It can hardly be proved that our world is contained in a ball, but it is also difficult to prove that it is not.

This chapter contains examples of somewhat unconventional approaches, and more abstract modeling tasks. The main purpose is to show models that perhaps have little practical applications, but offer new and cognitive topics and may stimulate our imagination.

In this section, we define a universe encapsulated in an open ball and simulate particle movements in it. The metrics and the linear vector space inside the ball are defined. They are based on the addition and multiplication operators defined below. Properties of this space are discussed and some simulations of a moving particles are described. This includes a simulation of the movement of several thousand particles that explode, starting from the initial singularity (a "small bang"). The four-dimensional space-time ball universe is defined. This ball is used to encapsulate the whole time-evolution of the three-dimensional world. Then, we extend the dimensionality of the ball, adding the fifth dimension, being a new, "hyper-time" coordinate. It is suggested that this process can be continued, by treating the time as a multi-dimensional variable.

### Nomenclature

$R_n$—n-dimensional real Euclidean space. We denote $R = R^3$

$B_n$—n-dimensional open ball with radius equal to one. We denote it as $B$ when the dimensionality is defined by the context.

$\hat{x} = (x, y, z)$—a point in $B$

$\hat{X} = (X, Y, Z)$—a point in $R$

$t$—the time

$d(\hat{u}, \hat{v})$—distance in $B$

$d(\hat{U}, \hat{V})$—distance in $R$

$M(\hat{x})$—a mapping $B \rightarrow R$

$M^{-1}(\hat{X})$—a mapping $R \rightarrow B$

$\oplus$—addition operator for two vectors in $B$

$\otimes$—multiplication operator (number by vector)

**NOTE:** In the following, the term *boundary of B* means the boundary of the closure of $B$.

## 12.6.1  General Remarks

This chapter is not related to any cosmological theory. We focus rather on the geometrical properties of the ball and the possible spatial structures defined inside.

The idea of a spherical universe is not new. However, it seems that it gains popularity in recent years with new observations and ideas. Di Valentino and Melchiorri [5] discuss an enhanced lensing amplitude in the cosmic microwave background power spectra, and point out that the closed Universe can provide a physical explanation for the effect. Linde [11, 12] and Efstathiou [6] discuss similar problems of a cosmic spatial curvature. Guth and Nomura [7] study the problem of the spatial curvature and its consequences.

The problem of symmetry breaking in the early history of the universe is discussed in Albrecht [2]. Aghanim et al. [1] describe a hybrid method with simulations based on the use of the high-frequency polarization data instrument and cosmic microwave background data.

There are a lot of related works in the available literature. We will not provide here a wider survey because the present paper is not exactly in the field of observations and their cosmological consequences. For more works on that field and more references, the reader may consult Hildebrandt et al. [8], Riess et al. [18], Sintunavarat [19] or Uzan [20]. For more information on the expanding universe and the theory of the "Big Bang" consult Peebles et al. [14].

Observe that if we admit a Universe to be enclosed in a limited ball or other regions, we should establish some kind of mathematical structure in it. From the mathematical point of view, the "Space of the Universe" should have a linear vector and metric structure. So, we define such structures inside an open n-dimensional ball with a radius equal to one. This permits us to use the operators of addition and multiplication of vectors, and to simulate the movement of sets of particles in our abstract ball-world.

## 12.6.2  The Ball

Imagine an abstract Universe that consists of a set of particles moving according to certain laws of physics, valid inside the open unit ball $B$. Recall that, as stated before, the term *boundary of $B$* means the boundary of the closure of $B$.

Before considering the events in $B$, we should define a space inside the ball. Up to now, $B$ is just a set of points (or vectors). To convert this set into space, we must define the certain structure(s) in it. Let's start with a metric structure. First, consider the mapping $M$ from $B$ to the real, three-dimensional Euclidean space $R$, defined as follows:

$$M(x, y, z) = (X, Y, Z) \text{ such that } \begin{cases} X = x/s, \ Y = y/s, \ Z = z/s, \\ \text{for } (x, y, z) \in B, \ (Z, Y, Z) \in R, \\ \text{where } s = 1 - x^2 - y^2 - z^2 \end{cases} \quad (12.15)$$

The inverse mapping is $M^{-1} : P \to B$, due to the formula

$$x = Xs, \ y = Ys, \ z = Zs, \quad (12.16)$$

where $s$ is defined in Eq. 12.15. In the following, the vector $-\hat{x} = (-x, -y. - z)$.

The algorithm for $M^{-1}$ is not so straightforward as that of $M$. To calculate $x$, $y$ and $z$, let sum the respective squared left- and right-hand sides of (12.16). We have

$$\begin{cases} x^2 + y^2 + z^2 = (X^2 + Y^2 + Z^2)s^2, \text{ where } s = (1 - x^2 - y^2 - z^2) \\ \text{thus, } 1 - s = (X^2 + Y^2 + Z^2)s^2 \\ (X^2 + Y^2 + Z^2)s^2 + s - 1 = 0 \end{cases} \quad (12.17)$$

Let $a = (X^2 + Y^2 + Z^2)$, $b = 1$ and $c = -1$.
This way, $s$ is a solution to the equation $as^2 + bs + c = 0$. So, we have

$$s = \frac{-b + \sqrt{b^2 - 4ac}}{2a} \quad (12.18)$$

Figure 12.13 shows the image of a uniform grid of points in $R$, mapped to $B$ using $M^{-1}$ (an intersection with the $x$, $y$ plane, with $z = 0$). Note that if a point in $R$ tends to infinity, then the corresponding point in $B$ approaches the boundary of the ball. Look at the section $x$ in Fig. 12.13. If such an object moves in the direction of the arrow, then its (absolute) length decreases. This means that the individuals that live near the limits of $B$ are smaller than these living near the center. However, they cannot notice this. The length of $X$ in terms of the distance $d$ (Sect. 12.6.3) is unchanged, and they have no reference to assess their real size.

It is interesting to see what happens if we accept the other, negative solution in (12.17). To do this, we can modify the mapping $M$, considering its domain as the

**Fig. 12.13** Mapping a
uniform grid of points from
R to B

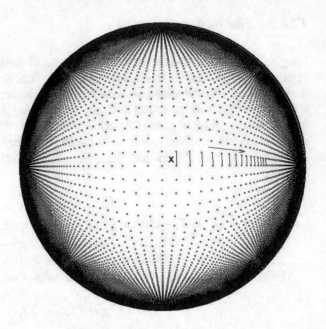

outside of $B$, namely $E = (cl(B))C$ (the complement of the closure of $B$). In other words, we suppose that $x^2 + y^2 + z^2$ in (12.17) is greater than one, and we take the negative solution in (12.18). Figure 12.14 shows the map of a regular grid in $R$ to the set $E$. Again, the infinity of $R$ maps to the boundary of $E$. The point "zero" (origin or $R$) maps into the infinity of $E$. In $E$, we can define a metric structure similar to that of the metric in $B$. The world created in $E$ will be called the *dual world* with respect to $B$. However, no particle that moves in $B$ can "jump" to the dual world, and vice versa, unless we admit some uncertainty in the mappings $M$ and $M^{-1}$ (see Sect. 12.6.9). In the following, we focus on the space $B$, rather than $E$.

Both $M$ and $M^{-1}$ are one-to-one mappings. Observe also that

$$M(0, 0, 0) = (0, 0, 0), \quad M^{-1}(M(\hat{a})) = (\hat{a}), \quad \text{and} \quad M(-\hat{a}) = -M(\hat{a}). \quad (12.19)$$

### 12.6.3  The Metric Structure

Let $D(\hat{U}, \hat{V})$ be a conventional distance between points in $R$. We define the distance $d$ in $B$ as follows:

$$d(\hat{u}, \hat{v}) = D(M(\hat{u}), M(\hat{v})) \ \forall \hat{u}, \hat{v} \in B \quad (12.20)$$

As $D$ satisfies the axioms of a metric, $d$ also does.

Absolute value or *norm* of $u \in B$ is $|u| = d((0, 0, 0), u)$ (defined in terms of $d(*)$ and not $D(*)$). We also denote $\| u \| = D((0, 0, 0), u) = \sqrt{u_x^2 + u_y^2 + u_z^2}$.

**Fig. 12.14**  The dual world

For example, let $u = (0.9, 0, 0)$. Then $\| u \| = 0.9$ and $|u| = 10.752$.

If we define a section of a straight line between two points $F$ and $G$ as the trajectory of minimal length that connects $F$ and $G$, then the straight lines in the ball $B$ with metrics $d$ become curves (*d-straight lines*). See some examples of d-straight lines in Fig. 12.15. A particle with a given initial velocity and no forces applied to it moves along such the d-straight lines (uniform line movement).

The simple Euclidean distance $D$ can be also calculated for two changes between-minus points in $B$, as $D(\hat{a}, \hat{b})$.

### 12.6.4   Linear Vector Space Operators

Vector sum in $B$:

$$
\begin{cases}
Z = (x_3, y_3, z_3) = (x_1, y_1, z_1) \oplus (x_2, y_2, z_2) \\
iff \\
(x_3.y_3, z_3) = M^{-1}(M(x_1, y_1, z_1) + M(x_2, y_2, z_2))
\end{cases}
\tag{12.21}
$$

Obviously, $(x_3.y_3, z_3) \in B$.

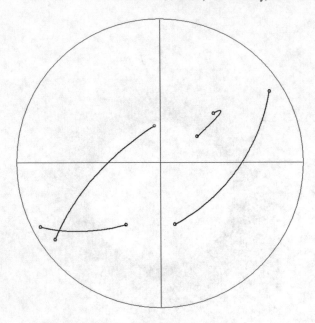

**Fig. 12.15** D-straight lines in the ball B, projection on the x-y plane

Now, calculate the following:

$\hat{a} \oplus (\hat{b} \oplus \hat{c}) = M^{-1}(M\hat{a}) + M(\hat{b} \oplus \hat{c})) =$
$M^{-1}(M(\hat{a}) + M(M^{-1}(M(\hat{b}) + M(\hat{c})))) = M^{-1}(M(\hat{a}) + M(\hat{b}) + M(\hat{c})),$

where $\hat{a}, \hat{b}, \hat{c} \in B$.

Calculating, in the similar way, the sum $(\hat{a} \oplus \hat{b}) \oplus \hat{c}$ we obtain the same result. This means that the summation operation $\oplus$ in $S$ is associative. Other axioms of the summation operator are also satisfied. The operator of multiplication is defined as follows ($K$ and $N$ are numbers, $M$ is the mapping (12.15)).

$$K \otimes \hat{a} = M^{-1}(KM(\hat{a})), \ K \in R^1 \tag{12.22}$$

We have (multiplicative axiom):

$K \otimes (N \otimes \hat{a}) = M^{-1}(KM(N \otimes \hat{a})) =$
$M^{-1}(KM(M^{-1}(NM(\hat{a}))) = M^{-1}(KNM(\hat{a})) = KN \otimes \hat{a}$

and distributive axioms (C, D–real numbers)

$C \otimes (\hat{a} \oplus \hat{b}) = C \otimes M^{-1}(M(\hat{a}) + M(\hat{b})) =$
$M^{-1}(CM(M^{-1}(M(\hat{a}) + M(\hat{b})))) =$
$M^{-1}(C(M(\hat{a}) + M(\hat{b})))$

$(C \otimes \hat{a}) \oplus (C \otimes \hat{b}) = (M^{-1}(CM(\hat{a})) \oplus (M^{-1}(CM(\hat{b})) =$

$$M^{-1}(CM(\hat{a}) + CM(\hat{b})) = C \otimes (\hat{a} \oplus \hat{b})$$

This way, we convert the set of all points in $B$ into a real linear vector metric space $B$. Let us consider the velocity of movement and use differential equations in $B$. By a *small vector* we mean a vector $\hat{a} \in B$, such that $\| \hat{a} \| << 1$. More precisely, we assume that in the mapping $M$ the factor $s$ (see (12.15)) may be supposed equal to one. So, for small vectors, mappings $M$ and $M^{-1}$ become identities. For small vectors we have

$(\hat{a} \oplus \hat{b}) = \hat{a} + \hat{b}, \quad k \otimes \hat{a} = k\hat{a}$

Note also that $\hat{a} \oplus (-a + r)$ becomes a small vector when $r$ approaches zero.

### 12.6.5  Local Ball and Local Observer

*Local ball* $LB$ is a ball included in $B$ with the diameter sufficiently small to neglect the changes of the value of the coefficient $s$ of (12.15) inside $LB$ ($s$ treated as a constant). Let's recall that here, we not only discuss the mappings between the $R^3$ Euclidean space and the interior of the ball. We suppose that **the real, physical world is located inside the ball, and not outside**.

A local observer is that who can get information only from the local sphere where he/she is located. The local observer concludes that he/she is leaving in the real linear vector space ($s$ is a constant, $M$ and $M^{-1}$ are linear mappings). For the local observer, the operations $\oplus$ and $\otimes$ reduce to the conventional operations on vectors in Euclidean space, and transformations $M$ and $M^{-1}$ become linear transformations. The local observer cannot know if the origin of his/her local world is located in the center of $B$ or not. In fact, he/she has no information about where his $LB$ is located.

Note that there is a simple transformation (self-mapping) that moves the origin of the $LB$ to the origin of $B$, defined as follows:

$$\hat{b} = \hat{a} \oplus (-\hat{a}_0)\forall \hat{a} \in B,$$

where $\hat{a}_0$ is the center of the $LB$

The above transformation is a mapping that maps the inside of the ball $B$ into itself, preserving all properties of the space $B$.

Let the spatial coordinates of a point $\hat{a}(t) = (x(t), y(t), z(t)) \in B$ depend on the time $t$. Denote $f(t) = (x(t), y(t), z(t))$. This describes a curve in $B$. We will name it the trajectory of $f(t)$. It is the graph of the function $(x(t), y(t), z(t))$ into the three-dimensional space $B$. If we define the $B$-world as a set of moving particles, then the set of the trajectories of all particles will be called *world trajectory*. Observe that to represent the state of a moving particle, we need its position, as well as the velocity. It could be supposed that the trajectory (recorded movement) already includes this information. However, to numerically retrieve the velocity from the movement, we would need to perform numerical differentiation. This operation is not recommended and may provide considerable errors. So, if we need to record the complete information about the moving particles, we must use six and not three variables for each point. This will be the world state trajectory. In this case, the spatial part of our world must be $B6$ ball instead the $B3$. In more general case, we may treat with a model of a complex dynamic system, which state is a point in an n-dimensional space. In this case, the spatial part of the model will be $Bn$-world, and the time will be treated as the coordinate $n + 1$. Section 12.6.10 provides more discussion on the 4th dimension of the $B$-world.

### 12.6.6   Velocity Superposition

Let $\hat{a}(t) = (x(t), y(t), z(t))$. Consider the expression

$$\hat{v}_a = \frac{d\hat{a}}{dt}|_B = \lim_{h \to 0} \left(\frac{1}{h}\right) \otimes \left[\hat{a}(t) \oplus (-\hat{a}(t - h))\right] \tag{12.23}$$

If the limit exists, then the above expression is the velocity of the point with respect to the origin of $B$. The subscript $B$ means that we use the operators defined in $B$.

Some authors admit that the real time advances in discrete time steps (a quantum-mechanics approach) [9]. If we admit that the time advances in finite time steps $\bar{h}$, then the formula (12.23) remains valid with only one change, namely without the "lim" operation. The discrete time case fits better with Heisenberg's uncertainty principle. According to this principle, we cannot assess both position and momentum of a particle simultaneously, basing on the particle state at one time instant. In the formula with finite $h = \bar{h}$, we use time instants $t$ and $t - \bar{h}$ to assess the velocity.

Velocity vector belongs to $B$. The superposition of vectors, for example, the velocities, in $B$ is different from the vector sum in $R$. For example, the superposition of two velocities in the $x$ direction, each of them equal to 0.2 (in terms of distance $D$), is the vector $0.2 \oplus 0.2$, which $D$ length is equal to 0.362.

With discrete time, the formula for the next position of a particle with a given velocity is the same as provided by the simple Euler's method: $\hat{a}(t + \bar{h}) = \hat{a}(t) \oplus (\bar{h} \otimes \hat{v}(t))$.

Recall that $|u| = d((0, 0, 0), u)$. We denote $\| u \| = D((0, 0, 0), u)$.

Consider a velocity $\hat{c}$, such that $\| \hat{c} \| = 1$. This vector does not belong to (open) $B$, and no particle can move with such velocity. Instead, we will consider a vector $\hat{c}_\varepsilon$, such that

$$\| \hat{c}_\varepsilon \| = \| \hat{c} \| - \varepsilon, \text{ where } \varepsilon > 0 \text{ and } \varepsilon << 1. \tag{12.24}$$

Velocity $\hat{c}$ is the limit case of $\hat{c}_\varepsilon$, when $\varepsilon$ approaches zero. Further on, we will discuss the velocity $\hat{c}_\varepsilon$ also referred to as the $\varepsilon$-velocity, assuming that $\varepsilon$ is an arbitrarily small parameter. However, we will not pass to zero with $\varepsilon$. Any velocity vector that satisfies (12.24) will be named $\varepsilon$-velocity.

Consider the sum $\hat{s} = \hat{v} \oplus \hat{c}_\varepsilon$.

We have

$\hat{v} \oplus \hat{c}_\varepsilon = M^{-1}(M(\hat{v}) + M(\hat{c}_\varepsilon))$,

but, $M(\hat{v}) + M(\hat{c}_\varepsilon) \approx M(\hat{c}_\varepsilon)$ because $M(\hat{c}_\varepsilon)$ tends to $\infty$ for small $\varepsilon$.

So, $\hat{v} \oplus \hat{c}_\varepsilon \approx M^{-1}(M(\hat{c}_\varepsilon)) = \hat{c}_\varepsilon$.

In other words, the sum of the $\varepsilon$-velocity and any other velocity in B is also an $\varepsilon$-velocity. This means that if a particle moves with velocity $\hat{v}$ and generates any object that moves with the $\varepsilon$-velocity, the velocity of that object is equal to the $\varepsilon$-velocity and **does not depend on** $\hat{v}$. Consider the product of $\varepsilon$-velocity and a real constant $k$.

$k \otimes \hat{c}_\varepsilon = M^{-1}(kM(\hat{c}_\varepsilon))$, $\hat{c}_\varepsilon = (c_{x\varepsilon}, c_{y\varepsilon}, c_{z\varepsilon})$, $\parallel \hat{c}_\varepsilon \parallel = 1 - \varepsilon$.

Denote $M(\hat{c}_\varepsilon) = \hat{d} = (d_{x\varepsilon}, d_{y\varepsilon}, d_{z\varepsilon}) \in R$.

Here, $k$ is a finite constant, $k >> \varepsilon$. As $\parallel \hat{c}_\varepsilon \parallel$ approaches 1 for small $\varepsilon$, the length of $D$ approaches infinity, and so does $kD$. According to (12.16), the infinity of $R$ maps into the boundary of $B$. This means that $|k \otimes \hat{c}_\varepsilon|$ is also the $\varepsilon$-velocity. A $\oplus$ sum of two or more $\varepsilon$-velocities in the same direction, is also an $\varepsilon$-velocity.

The constant $\varepsilon$ can be easily obtained from the quantum theory. Suppose that both time and space are discrete and that no particle or piece of information can "jump" more than one space step in one time step. Using the Planck minimal length equal to $1.616 \times 10^{-35}$ and the Planck minimal time-step equal to $t_m = 5.39 \times 10^{-44}$ we calculate the maximal speed in $R$ as $x_m = 299.8 \times 10^6$ m/sec. Calculating $M^{-1}(x_m)$, we obtain $\varepsilon \simeq 1.669 \times 10^{-9}$. However, this is only the estimate of the lower limit for $\varepsilon$, obtained from the time-space discretization, related to the speed of light. We do not discuss here the speed of light or any other issues that arise from the theory of relativity, focusing rather on the geometrical properties of the ball and not on the advanced mechanics.

### 12.6.7 Particle Movement and a Small Bang

Consider a set of particles (or material points) in $B$. Define the state of this set (the $B$-world) as the set of the positions and velocities of all particles. The *evolution* of the $B$-world is the set of the trajectories of all particles over the whole time interval $[-\infty, \infty]$. As $B$ is a normed linear vector space, we can suppose that the particle movement obeys the known equations of movement, expressed in terms of the operators $\oplus$, $\otimes$. In our simulations, the equations of movement are carried out using these operators. The same result can be obtained simulating the particle movement in $R$ (with conventional Euclidean geometry), and then, mapping the particle positions into $B$. However, we use the operators $\oplus$, $\otimes$ and the geometry of $B$ for a more conceptual reason: Our **real universe is inside, and not outside** the ball. The "external" space $R$ is treated as an auxiliary object, only

As an example, consider a small bang that occurs in $B$, over the time interval $[0, \infty]$. The force between two particles depends on the distance $d$ between them. Suppose that for small $d$ the particles repel, and for greater $d$ they attract each other. A simple expression of this force may be, for example,

$$f = g \left( \frac{1}{d^2} - \frac{a}{d^3} \right), \tag{12.25}$$

where $g$ is a constant, $a$ represents the "repelling radius" and $d$ is the distance between the two particles. Moreover, we suppose that the force between two particles that are located at the same point is equal to zero. Let's simulate the movement of $N = 5000$ particles.

The scenario for the small bang is as follows. In the beginning, all the particles are located in the same point $x_0$, $y_0$, $z_0$, with zero velocity (a singularity). Consequently, all the forces between particles are equal to zero and the system remains at a standstill. To an observer, the cloud of particles may even be noted as one, still particle. However, there is a huge potential energy in the system, that may be released. To release this energy, we apply a very small impulse of force to the particles. They start to move out of the singularity, receive a great force pulse of repelling force, and the explosion begins.

The integration of movement in a near neighborhood of the singularity is difficult. In our case, the repelling force is truncated to avoid numerical problems. We call this simulation "Small Bang." This is an implementation on a standard PC. Obviously, the simulation of the real Big Bang should be (and it is) rather carried out on a supercomputer.

In Fig. 12.16 we can see the images of some consecutive system states (the evolution of our $B$-world), marked as A, B, C, and D, respectively. First, the cloud of points rapidly expands. The initial repelling pulse results in an initial velocity that influences all the further expansion. The model is highly sensitive with respect to the parameters of the force expression. If the initial repelling is big, then the cloud expands to infinity, and no interesting structures are formed. On the other hand, for a smaller initial pulse, the attracting forces prevail and the cloud collapses one or more times. While collapsing, the points rapidly approach the initial singularity, and the forces between them grow. This results in a secondary explosion. In fact, the resulting velocities are the consequence of the imperfection of numerical integration, rather than the true particle movement. However, there are some parameter settings that result in the slow expansion after the initial "bang." If this slow expansion in reached, some local clusters of particles begin to appear, consolidate, and then remain in the slowly expanding cloud.

In the image of the final state (D) of the universe we can see the quite stable distribution of particles, with several clusters or "galaxies" (some of them rotating). It can be observed that the clusters are formed at the early stage of the evolution, but not just after the initial bang.

The force expression (12.25) for model of Fig. 12.16 has the following parameters: $g = 0.08$, $a = 0.01$, lower force limit $= -360$. The integration step was fixed to $h = 3 \times 10^{-7}$.

It should be noted that we can start simulating the particle movement in R, and then map the system state into the ball. This will provide the same images. However, this is not conceptually correct. *Our world is defined in B and not in R.* So, we should

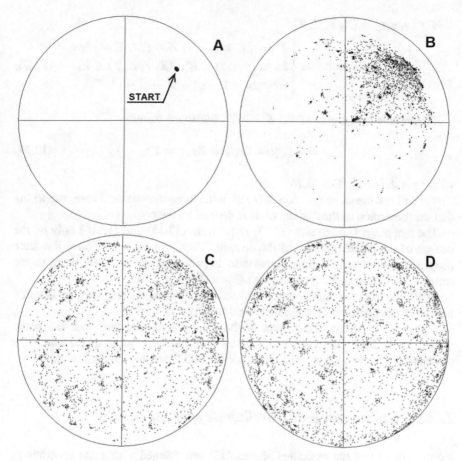

**Fig. 12.16** The expansion in the $B$-world; small bang. Projection on the x-y plane

simulate anything using the operators $\oplus$ and $\otimes$. This, of course, results in slower simulations.

## 12.6.8 *Adding the Time Dimension*

It is common to treat the time $t$ as a 4th dimension and consider events in the four-dimensional space with coordinates $(x, y, z, t)$. Up to now, $B$ was a three-dimensional open ball with radius one. Now, we add to it the time $t$ as the fourth dimension, the resulting ball being $B^4$. For this ball we redefine the mappings $M$ and $M_{-1}$ given by (12.15) and (12.16), respectively, as follows:

$$M(x, y, z, t) = (X, Y, Z, T)$$

$$\times \text{ such that } \begin{cases} X = x/s, \ Y = y/s, \ Z = z/s, \ T = t/s \\ \text{for } (x, y, z, t) \in B^4, \ (Z, Y, Z, T) \in R, \\ \text{where } s = 1 - x^2 - y^2 - z^2 - t^2 \end{cases} \quad (12.26)$$

The inverse mapping is $M^{-1} : R^4 \to B^4$, due to the formula

$$x = Xs, \ y = Ys, \ z = Zs, \ t = Ts, \quad (12.27)$$

where $s$ is defined in Eq. 12.26.

We will use the same notation $M(*)$ as in the three-dimensional case, supposing that the dimension of the domain of $M$ is defined by the context.

The mappings (12.26) and (12.27) differ from (12.15) and (12.16) only by the change of the dimensionality of the domain. They, as well as $B^4$, have the same properties as in the three-dimensional case. In particular, $B^4$ is a real, linear, vector metric space, equipped with operations $\oplus$ and $\otimes$.

In $B^4$, the time $t$ changes between minus and plus infinity. This way, we obtain a ball $B^4$ that incapsulates the **whole evolution** of our $B$-world.

Figure 12.17 shows the section of the four-dimensional ball with the plane $y = 0, z = 0$. Each curve is a set of points with the same value of the *time*. The constant time curves are shown both inside the ball B and for the dual world.


### 12.6.9  Uncertainty and Traveling Beyond the Infinity

Now, suppose that the mappings $M$ and $M^{-1}$ are charged with some uncertainty. This means that the result of the operations $\oplus$ and $\otimes$ provide *sets of possible values*, instead of a single point in $B$. Consequently, the particle movement and the evolution of the $B$-world become irreversible. Let start with a given particle position $\hat{a}(t)$ and velocity $\hat{v}(t)$. We have

$$\hat{a}(t + h) \in \hat{a}(t) \oplus (h \otimes \hat{v}(t)) \quad (12.28)$$

Due to the uncertainty, the right-hand side of (12.15) is a set and not a single point. Now, if we intend to go back in time (reversing the direction of the velocity), we must start from a point inside the set $U$ (Fig. 12.18) given by the right-hand side of (12.26). As the result, we obtain a set $W$ of possible positions at time instant $t$. This set may include or not the original point. This means that the particle can occupy two or more positions in the same time instant (Fig. 12.18), shown also in Sect. 12.5. If we don't permit this, the conclusion is that going back in time is impossible in our $B$-world. Note that in the case of uncertainty, the particle movement is described by a differential inclusion instead of a differential equation, and the movement is irreversible. This was also discussed in Sect. 12.5.

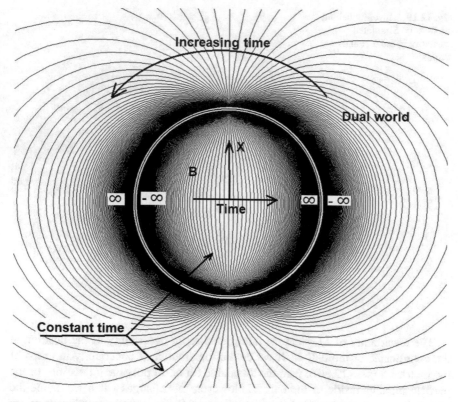

**Fig. 12.17** Time discretization, plane $t - x$. Curves of constant time

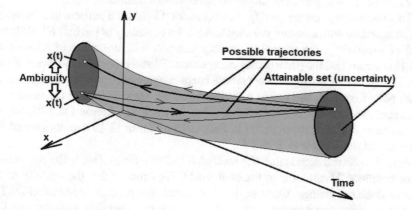

**Fig. 12.18** Going forward and back in time with uncertainty

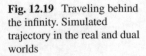

**Fig. 12.19** Traveling behind the infinity. Simulated trajectory in the real and dual worlds

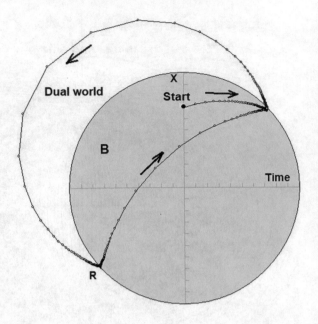

The uncertainty of $M^{-1}$ can be interpreted as an uncertainty of the position in the coordinates of the ball $B$. If we define the amount of uncertainty in terms of the metrics $d$ (see (12.20)), then the attainable set of all possible trajectories of the particle remains inside of the ball $B$. However, this changes when we define the uncertainty using the (simple Euclidean) distance $D$. The source of such uncertainty may also be a consequence of time- and space discretization.

This uncertainty has interesting consequences. Consider a particle that moves on a $d$-straight line with constant velocity, toward the boundary of the ball $B^4$. With this kind of uncertainty, the attainable set may intersect with boundary of (closure of) $B^4$. This means that the particle may escape from $B^4$ and enter the dual world $E$ (see Fig. 12.17). Once entering $E$, the particle keeps moving according to the kinematics of $E$. Note that in $E$, the vector operations and the metrics remain the same as in $B$, except the formula for $M^{-1}$ (the negative solution in taken in (12.17)). So, we can simulate the further movement of the particle. Figure 12.19 depicts a simulated trajectory of the particle in $B$ and $E$.

First, the particle approaches the limit of $B$ (infinite time). Due to the uncertainty in the positions, it can jump to the dual world. We suppose that the velocity of the particle does not change. Once in the dual world, the particle keeps moving. The simulation shows that it approaches the limit of $B$ at the other side, with time tending to infinity (in the dual world). Then, due to the same uncertainty, the particle can jump to the ball $B$ again, and keep moving. This cycle is repeated infinite times. Note that the trajectory in the dual world is not completely defined because of the uncertainty of particle position just after the first jump. So, after returning to $B$, the trajectory may, or not, pass through the original starting point. However, the reentry point $R$ is always the same for each possible trajectory.

## 12.6.10  The Fifth Dimension

In $B^4$, the time $t$ changes between minus and plus infinity. This way, $B^4$ encapsulates the whole evolution of our $B$-world.

Now, the question is why our space is limited to four dimensions? Our mind has been designed to understand the three-dimensional environment and the time. However, this is the only reason to limit us to the four-dimensional space-time universe. In other words, this limitation is strictly subjective. Thus, consider an extension of our universe to n-dimensional space. Let us consider an additional, fifth dimension with the corresponding coordinate $u$. Transformations (12.26) and (12.27) remain similar, only the coordinate $u$ (a "hyper-time") is added, so that

$$s = 1 - x^2 - y^2 - z^2 - t^2 - u^2$$

The coordinate $u$ is a "hyper-time." For $B^5$, we can define new laws of physics. We can use the same principles of causality and uncertainty as in $B^4$, or define completely different rules. Remember that here, we are creating the world(s) and we don't look for analogies with the real world (this is left to the reader's imagination).

The new $B^5$-universe encapsulates the "evolution of evolutions." In other words, it encapsulates the changes of all the possible evolutions of $B^4$. Figure 12.20 illustrates the process of adding new dimensions to $B$. Each ball $B^3$ represents one state of the three-dimensional world. The series of all consecutive $B^3$s form the evolution that is encapsulated in $B^4$. In turn, advancing in the hyper-time $u$, we have a new "evolution of evolutions," that is enclosed in $B^5$. This process may be continued by adding new "hyper-time" coordinates. In fact, we should ask ourselves if the time must be a one-dimensional variable, and why not n-dimensional?

Though the fifth dimension and the two-dimensional time is difficult to accept by our intuition, this extension of the concept of time has some elements similar to the one-dimensional time. Consider a continuous dynamic system, with a one-dimensional state variable. With sufficient regularity assumptions, the system movement can be described by an ordinary differential equation

$$\frac{dx}{dt} = f(x, t), \text{ or, in a more general form, } f(x, x', t) = 0 \qquad (12.29)$$

Now, suppose that the time is two-dimensional, and each time instant is a point on the plane $(t, u)$ where $u$ is the other time coordinate. Now, the system dynamics may be described by a set of partial differential equations, instead of the ordinary one, as follows:

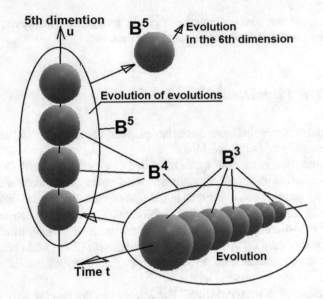

**Fig. 12.20** Adding the fifth and sixth dimension

$$
\begin{cases}
\dfrac{\partial x}{\partial t} = f_{11}(x, y, z, t, u), & \dfrac{\partial x}{\partial u} = f_{12}(x, y, z, t, u) \\[2mm]
\dfrac{\partial y}{\partial t} = f_{21}(x, y, z, t, u), & \dfrac{\partial y}{\partial u} = f_{22}(x, y, z, t, u) \\[2mm]
\dfrac{\partial z}{\partial t} = f_{31}(x, y, z, t, u), & \dfrac{\partial z}{\partial u} = f_{32}(x, y, z, t, u)
\end{cases} \tag{12.30}
$$

## 12.6.11  Conclusion

In this chapter, a non-conventional application of differential inclusions is discussed. In this application, the problem is if we can integrate system trajectories with ideal predictor ("feedback from the future"). The uncertainty is defined as the set of all possible future model states. This permits us to calculate the reachable set. It is pointed out that the iterative procedure described here may converge to the unique trajectory that is the solution to the ideal predictor problem.

Other parts of the chapter deal with the idea of encapsulating our Universe in an open n-dimensional ball of radius one. We can define a metric and linear vector structure inside the ball. This makes it possible to discuss particle movements inside the ball using ordinary differential equations. There is one-to-one mapping between the ball space and the space $R^n$. The main tenet is that *the real world is enclosed in the ball, and the Euclidean space R around is only an additional, abstract structure*. Adding the fourth dimension that represent the time, we can encapsulate the whole evolution of our *B*-world into the ball. Then, we can consider the fifth and higher dimensions, adding a "hyper-time" coordinates, and encapsulate "evolution of evolutions" in similar balls, with higher dimensions. If we add an uncertainty to the mapping from and into the ball, it is possible that a moving particle goes through the limit of B. The particle keeps traveling behind the infinity and enters a "dual world" defined outside the ball.

# References

1. Aghanim N, Akrami Y, Arroja F, Ashdown M (2019) Planck 2018 results. Cosmological parameters. Astronomy & Astrophysics, Cornell University, VI. https://doi.org/10.1051/0004-6361/201833880
2. Albrecht A, Steinhardt PJ (1982) Cosmology for grand unified theories with radiatively induced symmetry breaking. Phys Rev Lett 48:1220–1223
3. Aubin JP, Cellina A (1984) Differential inclusions. Springer Verlag, Berlin
4. Aubin JP (2013) Tychastic viability?: a mathematical approach to time and uncertainty. 61(3):329–340. https://doi.org/10.1007/s10441-013-9194-4
5. Di Valentino E, Melchiorri A, Silk J (2019) Planck evidence for a closed Universe and a possible crisis for cosmology. Nat Astron 4:196–204. https://doi.org/10.1038/s41550-019-0906-9
6. Efstathiou G (2003) Is the low cosmic microwave background quadrupole a signature of spatial curvature? Mon Not Roy Astron 343:L95–L98
7. Guth AH, Nomura Y (2012) What can the observation of nonzero curvature tell us? Phys Rev D 86. https://doi.org/10.1103/PhysRevD.86.023534
8. Hildebrandt H, Viola M, Heymans CJ, Joudaki S (2017) Cosmological parameter constraints from tomographic weak gravitational lensing. 405(2), https://doi.org/10.1093/mnras/stw2805
9. Jaroszkiewicz G, Norton K (1997) Principles of discrete time mechanics: I. Particle systems. J Phys A: Math Gen 30(9):3115–3144. https://doi.org/10.1088/0305-4470/30/9/022
10. Lee EB, Markus L (1967) Foundations of optimal control theory. Wiley, New York, pp 978–0898748079
11. Linde AD (2003) Can we have inflation with O>1. Cosmol Astropart Phys 0305(002). https://doi.org/10.1088/1475-7516/2003/05/002.
12. Linde AD (1982) A new inflationary Universe scenario: a possible solution of the horizon, flatness, homogeneity, isotropy and primordial monopole problems. Phys Lett B 108:389–393. ISBN/ISSN 0370-2693
13. Nahin PJ (1998) Time machines: Time travel in physics, metaphysics, and science fiction. Aip Press, Springer. 0-387-98571-91998
14. Peebles PJE, Schramm DN, Turner EI, Kron RG (1994) The evolution of the universe. Scentific American 271(4):52–57
15. Pontryagin LS, Boltyanskii VG, Gamkrelidze RV, Mishchenko EF (1962) The mathematical theory of optimal processes. Interscience, ISBN: 2-88124-077-1
16. Raczynski S (2011) Uncertainty, dualism and inverse reachable sets. Int J Simul Model 10(1):38–45. ISBN/ISSN:1726-4529
17. Raczynski S (2002) Differential inclusion solver. Conference paper. In: International Conference on Grand Challenges for Modeling and Simulation, SCS, San Antonio TX
18. Riess AG et al (2018) New parallaxes of galactic cepheids from spatially scanning the hubble space telescope: implications for the Hubble constant. Astrophys. J. 855:136
19. Sintunavarat W, Cho YMJ (2012) Coupled fixed-point theorems for contraction mapping induced by cone ball-metric in partially ordered spaces. Fixed Point Theory and Applications, 128, Springer, https://doi.org/10.1186/1687-1812-2012-1
20. Uzan JP, Kirchner U, Ellis GFR (2003) Wilkinson microwave anisotropy probe data and the curvature of space. Mon Not Roy Astron Soc 344:L65–L68. https://doi.org/10.1046/j.1365-8711.2003.07043

# Index

**A**

Activity scanning, 174
Additivity, 5
Advertising, 160
Agent-based model, 174
Aggregation, 3
Airplane dynamics, 142
Analog computer, 25, 39
ANSYS, 68
Arena, 171
Artificial societies, 191
Attainable set, 83

**B**

Backward reachable set, 259
Basic model, 12
Big-bang (small), 269
Birth-and-death model, 16
Black box model, 5
Block diagram, 32
BLUESSS, 197
Bond graph, 32, 61
Buttlefield model, 223

**C**

Car suspension, 52
Category, 229, 231
Causal system, 27
Causality, 27, 238
Chicken game, 212
Collision of bodies, 231, 233
Competition model, 161
Complex models, 5
Computational complexity, 10
Computational sociology, 190

Computational tractability, 10
Concentrated parameter, 28
Continuous model, 25
Control system, 85
Corruption model, 192, 193
Coupled semi-discrete models, 217
Credibility, 17
CSL language, 174

**D**

Demand model, 159
Deterministic, 5
DEVS, 11, 177, 178
Differential inclusion, 5, 81, 83
Differential inclusion solver, 88, 110, 249
Discrete differential inclusion, 95
Discrete event, 210
Discrete event model, 10, 171
Discrete reachable set, 96
Distributed parameter, 5, 29, 65
Distributed simulation, 180
Diststrictly, 209
DYMOLA, 64
Dynamic optimization algorithm, 156

**E**

Economic growth model, 6
Economy dynamics, 6
Empirical model, 5
Encapsulated Universe, 261
End-user, 17
Equivalent inputs, 28
Event scheduling, 174
Experimental frame, 12, 20
Explicit model, 5

© The Editor(s) (if applicable) and The Author(s), under exclusive license to Springer
Nature Switzerland AG 2022
S. Raczynski, *Models for Research and Understanding*, Simulation Foundations,
Methods and Applications, https://doi.org/10.1007/978-3-031-11926-2

**F**
Farol model, 176
FCTS, 67
Feedback from the future, 251, 254
Fifth dimension, 275
Finite automata, 28
Finite element, 68
Finite-Time Event Specification, 211, 214
Flight dynamics, 142
Flight reachable sets, 144
Frequency response, 36
Functional sensitivity, 87, 108, 109
Functor, 230, 232
Fuzzy discrete event, 210
Fuzzy time instant, 238

**G**
GPSS, 11, 171

**H**
Hamiltonian, 152
Hausdorff distance, 207
Hedonic model, 161
Homogenity, 5
Human population model, 191

**I**
Ideal predictor, 254
Infinite automata, 28
Input-output validity, 14
Inverse reachable set, 259

**K**
Keynesian approach, 7

**L**
Lancaster model, 161
Landing on the Moon, 153
Laplace transform, 33
LAX mathod, 67
Linearity, 5, 29
Local observer, 267
Logarithmic gain, 107
Lotka–Volterra model, 111

**M**
Market elasticity, 159, 160
Market model, 130, 157

Market optimization, 151, 157, 163
Market optimization results, 163
Market utility, 160
Mason rule, 58
Matched pole-zro, 44
Mathematical model, 3
Maximum principle, 89, 151, 152
Maximum principle, iterative, 156
Model sensitivity, 107
Model state, 27
Modeling, 1, 3
Morfism, 229, 231
MSEIRS model, 8
Multi-dimensional time, 238

**N**
Navier-Stokes equation, 65
Non-linearity, 3
Numerical method, 31
Nyquist criterion, 38

**O**
Object-oriented model, 171
ODE, 30
ODE model, 5, 9, 52
Optimal control, 151
Optimal trajectory, 97
Ordinary differential equation, 5, 30
Organization model, 193
Organization theory, 190

**P**
Particle movement, 269
PDE model, 9
People agents, 176
Petri net, 11, 179
PID control, 121
Pole-zero mapping, 44
Political Map, 193
Predictor, 247
Probabilistic model, 5
Process interaction, 174
ProModel, 171

**Q**
Quasitrajectory, 84
Queuing model, 173

**R**
Rational actor, 176

Reachable set, 83, 94
Regression analysis, 107
Revenue (market model), 160
Reversibility, 258
Routh/Hurwitz criterion, 34


**S**
Sampled data system, 40
Seasonal index, 160
Selector function, 83
Self-organization, 193
Semi-discrete event, 211, 214
Sensitivity, 107
Signal flow diagram, 32
Signal flow graph, 56
Simulation, 1
Simultaneous events, 177, 233, 235
Singularity in the space of models, 224
SIR mosel, 8
SIRS model, 8
Sobol method, 107
Social system model, 189, 190
Soft systems, 3
Solow-Swan model, 7
Solution to tr DI, 83
Stability, 34
State events, 173
State of model, 27
Stiff equation, 51
Stochastic, 9
Stochastic model, 18
SWARM, 175
System, 3, 26
System classification, 28
System Dynamics, 16

**T**
Tendor set, 84
Three-phase strategy, 174
Time and event management, 179
Time arrow, 237
Time warp, 181
Time-shift, 248
Tractability, 10
Transfer function, 6, 32
Traveling beyond infinity, 272
Tychastic variable, 87


**U**
Uncertain future, 249, 251
Uncertainty, 17, 86, 247
Universe in a ball, 262
Universe madel, 261
Utility function, 160


**V**
V/f speed control, 115
Validity, 12, 13, 231, 233
Vehicle movement, 127


**W**
World evolution, 275


**Z**
Z-transform, 40